Solving tribology problems in rotating machines

Related titles:

Surface coatings for protection against wear
(ISBN-13: 978-1-85573-767-9; ISBN-10: 1-85573-767-1)
This authoritative book presents an overview of the current state of research in, and applications for, wear protective coatings. It concentrates on the different types of surface technologies which are used for protective coatings. Each chapter provides an in-depth analysis of a particular type of surface coating, its properties, strengths and weaknesses in various applications. Each surface coating treatment examined includes case studies describing its performance in a specific application. *Surface coatings for protection against wear* will be an invaluable reference resource for all engineers concerned with the latest developments in coatings technology.

Surface engineering casebook: Solutions to corrosion and wear-related failures
(ISBN-13: 978-1-85573-260-5; ISBN-10: 1-85573-260-2)
This book concisely and uniquely encompasses the principles of corrosion and wear as manifested in industrial failures and the solutions offered by surface engineering.

Engineering catastrophes: Causes and effects of major accidents, 3rd edn
(ISBN-13: 978-1-85573-505-7; ISBN-10: 1-85573-505-9)
This new edition contains a general update of historical data, more material concerning road and rail accidents and, most importantly, a new chapter on the human factor. The author provides a broad survey of the accidents to which engineering structures and vehicles may be subject. Historical records are analysed to determine how loss and fatality rates vary with time and these results are displayed in numerous graphs and tables. Notable catastrophes such as the sinking of the *Titanic* and the *Estonia* ferry disaster are described. Natural disasters are considered generally, with more detail in this edition on earthquake resistant buildings.

Details of these and other Woodhead Publishing materials books and journals, can be obtained by:

- visiting our web site at www.woodheadpublishing.com
- contacting Customer Services (e-mail: sales@woodhead-publishing.com; fax: +44 (0) 1223 893694; tel.: +44 (0) 1223 891358 ext. 30; address: Woodhead Publishing Ltd, Abington Hall, Abington, Cambridge CB1 6AH, England)

If you would like to receive information on forthcoming titles, please send your address details to: Francis Dodds (address, tel. and fax as above; email: francisd@woodhead-publishing.com). Please confirm which subject areas you are interested in.

Solving tribology problems in rotating machines

H. Prashad

CRC Press
Boca Raton Boston New York Washington, DC

WOODHEAD PUBLISHING LIMITED
Cambridge England

Published by Woodhead Publishing Limited, Abington Hall, Abington, Cambridge CB1 6AH, England
www.woodheadpublishing.com

Published in North America by CRC Press LLC, 6000 Broken Sound Parkway, NW, Suite 300, Boca Raton, FL 33487, USA

First published 2006, Woodhead Publishing Limited and CRC Press LLC

British Library Cataloguing in Publication Data
A catalogue record for this book is available from the British Library.

Library of Congress Cataloging in Publication Data
A catalog record for this book is available from the Library of Congress.

Woodhead Publishing ISBN-13: 978-1-84569-110-3 (book)
Woodhead Publishing ISBN-10: 1-84569-110-5 (book)
Woodhead Publishing ISBN-13: 978-1-84569-111-0 (e-book)
Woodhead Publishing ISBN-10: 1-84569-111-3 (e-book)
CRC Press ISBN-10: 0-8493-9209-8
CRC Press order number: WP9209

The publishers' policy is to use permanent paper from mills that operate a sustainable forestry policy, and which has been manufactured from pulp which is processed using acid-free and elementary chlorine-free practices. Furthermore, the publishers ensure that the text paper and cover board used have met acceptable environmental accreditation standards.

Typeset by Replika Press Pvt Ltd, India
Printed by TJ International Ltd, Padstow, Cornwall, England

Contents

Preface

This book presents a discussion of the solutions to various complex technical problems concerning bearings and lubricants, that have been diagnosed as closely associated with the design, quality, malfunctioning, maltreatment, and operation of machines. The various problems, their analyses and solutions discussed in this book are unique in nature, but the symptoms of the problems and related failure were projected to be closely concerned with tribology. Detailed diagnosis following repeated trials revealed the source of the problems and thereby the origin of the failures, which were dormant in the rotating machines. The solutions to these problems opened a new era and direction for development and analysis in the field of industrial tribology for tackling similar problems in power plant equipment.

The various chapters of this book deal with individual problems and their solutions, particularly pertaining to the progressive increase in bearing vibrations of a large synchronous electric motor, intermittent whistling noise from the bearings of vertical pump-motor sets, magnetization of bearings leading to premature failure in alternators being improperly located on the shaft, bearing failure of a synchronous condenser by the use of contaminated lubricant, bearing and design modification of windmills for trouble-free operation, unrecognized flow of a localized electric current in the rolling-element bearings causing flutings by the induction effects on the track surfaces of races, and various unforeseen causes leading to the bearing failure of large electric motors. Furthermore, techniques are shown for axial force measurement on a rotor to assess design and manufacturing accuracy of large electric motors, precise reliability assessment of quality and deterioration of rolling-element bearings in operation, effect of cage and roller slip on defect frequency response and functional performance tests of the rolling-element bearings to ensure their acceptance in routine applications in various industrial equipment.

The effects of viscosity, oil grades, clearance ratios, optimum values of design parameters of a spherical seating of hydrodynamic journal bearings and a regime of bearing operation have been analysed to provide trouble-free performance of the bearings in industrial equipment. The life expectancy of turbine oils, highlighting a methodology and criterion for their acceptance/rejection in power plant equipment for the reliable operation in the system, is very significant for maintenance engineers. A chapter on magnetic suspension bearings for energy meters emphasizes the importance of refinements that are necessary in the design and manufacturing of magnetic bearings. A chapter on state-of-the-art bearing response is also included, as this has potential in the analysis of the performance of a roller bearing operating under the influence of an electrical current.

Furthermore, the concept, development and investigation of the new generation double-decker high-precision rolling-element bearings with an outline of their performance advantages are included as a development approach through modulating bearing kinematics. The published literature on the evolution of dynamic coefficients and stress distribution on the outer surface of these bearings, their energy-efficient performance characteristics through experimental and theoretical investigations, including the effect of centrifugal forces and axial deflection, have been discussed to outline their advantages.

In short, this book deals with unique multifaceted problems and their solutions and provides in-depth analysis and investigations of bearings and lubricants and system problems, including reliability assessment and new bearing design, in 17 chapters. The book presents original typical case studies drawn from the author's professional experience over the past three decades in the area of industrial tribology. The work presented may prove to be useful for engineers and technologists of heavy industry, students, research engineers/scientists, academicians and others who grapple with the complex problems of industrial tribology. Various systems of units have been used in addition to SI to improve the projection and interpretation of results.

The author is extremely thankful to the large number of professionals/engineers and colleagues from various units of Bharat Heavy Electricals Ltd and from BHEL Corporate R&D Division, who have participated in different capacities in solving/rendering assistance to identify difficult and multi-dimensional complex industrial tribology problems closely linked with system design and operation. Thanks are due to BHEL, Corporate R&D management for providing an opportunity to handle in-depth involved systems and complex unique problems concerning industrial tribology during my professional career.

Furthermore, without the silent sacrifice of my wife Darshan, my beloved daughter Shwetlana, and my loving and caring son Poojan, who allowed me to work without any interference at a time when they needed my involvement the most, this task would not have been completed. Finally, my gratitude is due to the Almighty, whose blessings, radiance, continued inspiration, pouring of booming energy and silent guidance have nurtured me from time to time to follow the path of untiring research.

Dr Har Prashad

About the author

Dr Har Prashad is a Senior Deputy General Manager at Bharat Heavy Electricals Limited, Corporate Research and Development Division, Hyderabad. He obtained an ME (Hons) degree in Mechanical Engineering in 1970 and, subsequently, a PhD in Tribology. He worked with the Indian Institute of Petroleum, Dehra Dun, and with the Design Bureau at Bokaro Steel Ltd, Dhanbad, Bihar, before joining BHEL in 1974.

At BHEL, he was associated with the setting up of the Tribology Laboratory at the Corporate R&D Division, Hyderabad, particularly the design and development of various bearing test rigs for hydrodynamic and rolling-element bearings. He has done substantial design/development work on magnetic, dry and other special types of bearings. He has developed energy saving double decker high-precision rolling-element bearings, and established the performance of these bearings both theoretically and experimentally.

His areas of interest include diagnostic monitoring, failure analysis and bearing performance evaluation. He has done significant original work to establish the behaviour of different bearings and lubricants under the influence of electrical current. He has established electrical analogy for dynamic analysis of bearings. Assessment of flow of current through rolling-element bearings by study of magnetic flux density distribution on the bearing surfaces, theoretical evaluation of corrugation pattern, bearing life estimation, resistivity and recouping of resistivity phenomenon in lubricants, and theory that explains the causation, morphology and rate of formation of electrical current damage are some of the outstanding original contributions of Dr Prashad especially for engineers engaged in the design or operation of heavy rotating electrical machinery.

Dr Prashad has published more than 115 papers in both national and international journals, and delivered invited talks. He has patents to his credit.

He is a recipient of the Corps of Electrical and Mechanical Engineering Award–1998 along with other various awards for his contributions and publications.

Contact details:
Dr Har Prashad
1-2-319/A Gagan Mahal
Domal Guda
302 Central View Apartments
Hyderabad-500029
India

E-mail: har.prashad@gmail.com

(Ex Senior Deputy General Manager (Tribology), Bharat Heavy Electricals Limited (BHEL), Corporate Research and Development Division, Hyderabad, India.)

1
Reliability analysis of rolling-element bearings

1.1 A general review

Investigations employing the high-frequency resonance technique (HFRT), which diagnoses defect frequencies of rolling-element bearings of different makes, have been carried out. Raw vibration signatures of bearings at different speeds of operation have been demodulated, and envelope-detected spectra analysed to evaluate various defect frequencies and their energy levels. A relative comparison of various bearings has been made on the basis of identified defect frequencies and severity of defects. These frequency values and their energy levels are used to monitor the intrinsic conditions of bearings as well as to establish the severity of existing or developing defects in the bearings.

The investigation gives a realistic approach to monitor the intrinsic condition of a bearing. It can be successfully utilized to select appropriate bearings and for performance evaluation, and can act as a reliable tool to establish a safe bearing operational limit. Investigations can serve as a precise quality control instrument for the earliest detection of defects of even the smallest nature, and can be used as an 'on-line' bearing condition monitor, if required.

1.2 Introduction

Normal rolling-element bearings generate an easily identifiable ball or roller pass frequency when operating. Bearing defects amplify the amplitude of these frequencies. By using the high-frequency resonance technique (or envelope detection), these frequencies can be isolated and demodulated to give an indication of bearing condition.[1]

Defects in a bearing generally will produce impacts when in contact with mating parts. The secondary effect of this impact is to excite resonance in the races, rolling-elements or other structural elements. These high-frequency resonances decay exponentially and are modified (modulated) at ball/roller pass frequency in a manner that can be easily detected. The basic signal is obtained using a high-frequency accelerometer located as near the bearing as possible. The resulting spectrum is analysed to detect possible resonant frequencies. Once the frequency of interest has been found, the signal is narrow band filtered and the filtered output is envelope-detected. The spectrum of the envelope is obtained and analysed in the range of defect frequencies. The frequency of interest is generally some structural resonance, which can be found analytically or experimentally. Quite a few investigations have been reported using this technique.[2-5]

1.3 Detection of bearing malfunction and determination of defect frequencies

In general, the possible causes of bearing failure include excessive contact stresses, misaligned loads, material flaws, lubricant failure or contamination and electric discharge between rolling-elements and track surfaces of races. In any case, malfunction is manifested as a defect in a race or a rolling-element. Such malfunction can be detected by frequency analysis from a signal obtained from a bearing, based on a position of a given fault, i.e. race or ball/roller. Impacts and vibrations will be produced at frequencies that are functions of a component and speed of operation. For example, if a race is scratched, every time a rolling element makes contact with the scratch, an impact will be transmitted through a bearing. This impact will repeat itself as a function of bearing rotation. Spectral identification at these frequencies is the basis of many diagnostic systems. The defect frequencies can be derived from kinematics analysis of the defect impacting action of the rolling-element bearings.

Assuming that the inner race is rotating and that the outer race is stationary, the following formulae can be derived[6] (See Section 1.11 for nomenclature):

$$f_c = \frac{0.5 f_s}{1 - (d/D)\cos\alpha} \qquad [1.1]$$

$$f_b = 0.5 f_s (D/d)\,[1 - (d/D)^2 \cos^2\alpha] \qquad [1.2]$$

$$f_{bf} = 2nf_b = nf_s (D/d)\,[1 - (d/D)^2 \cos^2\alpha] \qquad [1.3]$$

$$f_{ORDFL} = nNf_c = \frac{0.5\, nNf_s}{1 - (d/D)\cos\alpha} \qquad [1.4]$$

$$f_{IRDFL} = nN(f_s - f_c) = \frac{0.5\, nNf_s}{1 + (d/D)\cos\alpha} \qquad [1.5]$$

$$f_{IRWL} = f_{IRDFL} \pm f_s \qquad [1.6]$$

$$f_{IRWNL} = \frac{f_{IRDFL}}{N \pm f_c} \qquad [1.7]$$

$$f_{\mathrm{ORDFNL}} = (n \pm 1) f_{\mathrm{c}} \qquad [1.8]$$

$$f_{\mathrm{RRS}} = 2nf_{\mathrm{b}} \qquad [1.9]$$

$$f_{\mathrm{RW}} = 2nf_{\mathrm{b}} \pm f_{\mathrm{c}} \qquad [1.10]$$

1.4 Resonant frequencies

Resonant frequencies can be initiated by shock loading. A defect in a rolling-element bearing may excite the resonance in an inner or outer race, or in the ball/roller. Because of the interaction, a combination of resonant frequencies may be excited, thus causing 'ringing', which is a continuous vibration response of the bearing structure. Both races and rolling-elements can exhibit this 'ringing'. It is characterized by an exponentially decaying, high-frequency oscillation. The 'ring' will appear periodically at the 'pass' frequency corresponding to the faults that excite it. Resonant frequencies are structural characteristics. Their frequency does not depend on the speed of rotation of the bearing.

The frequency of free resonance of a rolling-element may be calculated by the following formula:[7]

$$f_{\mathrm{br}} = \left(\frac{0.848}{2r}\right)\left(\frac{E}{2\rho}\right)^{0.5} \qquad [1.11]$$

The races will also exhibit resonant characteristics. Resonant frequencies for races can be calculated by the following relation [1.7]:

$$f_{\mathrm{rr}} = \frac{K(K^2 - 1)}{2\pi a^2 (K^2 + 1)^{0.5}} (EI/m)^{0.5} \qquad [1.12]$$

These resonant frequencies are in a free state and may alter when mounted on a structure.

1.5 Experimental procedure

The drive motor assembled with bearings under test was coupled to a similar motor through a rigid coupling. The test set-up is shown in Fig. 1.1. The coupled motor acts as a generator, while the drive motor, assembled with different kinds of bearings for investigation, acts as the motor under test. The speed of the motor was varied by changing the voltage and current input to the motor. The speed was monitored by a magnetic pick-up and displayed on a digital speed indicator. All the bearings were tested under coupled mode of operation for over-speed up to 2750 rpm up to maximum speed of 2275 rpm. Bearing geometrical parameters – bearing run-out, bearing clearance, swell, inner diameter of inner race and outer diameter of outer race – before and after assembly were also monitored in each test.

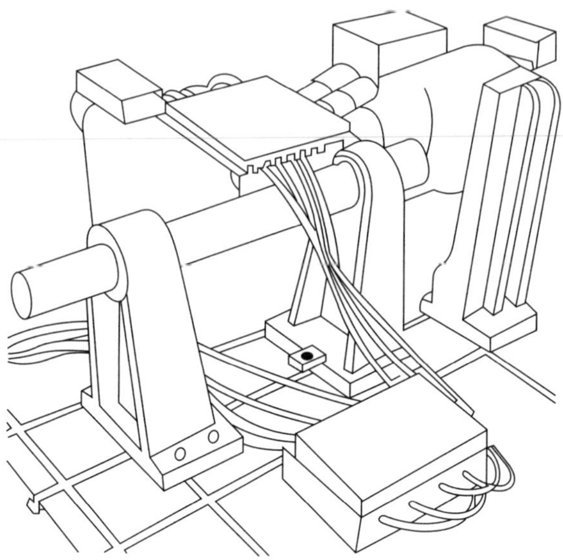

1.1 Test set-up.

1.6 Instrumentation details and techniques

1.6.1 Vibration signature and recording system

In all the tests, vibration accelerometers were mounted in four different locations, to record radial and axial vibration signals as shown in Fig. 1.2. Two pick-ups with magnetic bases were mounted radially, one on the outer race and the other on the end shield. In addition, one pick-up with a magnetic base was mounted axially on the end shield, while one pick-up with a stud base was mounted radially on the end shield to assess the attenuation of the signal. The signals from the different pick-ups were recorded on the four-channel B&K tape recorder in acceleration mode, to carry out further detailed analysis of the signatures by using a real-time analyser in the laboratory.

1.6.2 Investigations using the high-frequency resonance technique

A number of resonances can be detected by an accelerometer mounted on the bearing surface. The signal from accelerometer, after some amplification, is tape recorded for high-frequency resonance analysis at a later time. When setting up such an analysis system, the output signal from the accelerometer is directly analysed in the real-time frequency spectrum analyser. This step helps in determining the carrier (resonant) frequencies, which are being most strongly modulated by the defect frequencies. The accelerometer signals

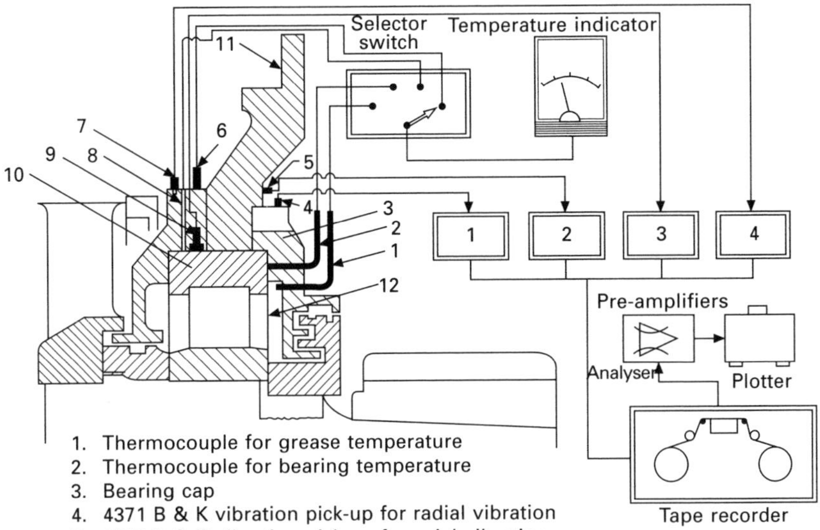

1. Thermocouple for grease temperature
2. Thermocouple for bearing temperature
3. Bearing cap
4. 4371 B & K vibration pick-up for radial vibration
5. 4371 B & K vibration pick-up for axial vibration
6. Thermocouple for end shield temperature
7. Stud base vibration pick-up
8. Spring-loaded thermocouple
9. Magnetic base vibration pick-up
10. Roller bearing
11. End shield
12. Grease space

1.2 Motor with instrumentation details and scheme.

are then envelope-detected about these frequencies in order to determine which of them is the best carrier for the particular defect frequency.

Envelope detection system

The system comprises an input amplifier, a multiplier, a tuned filter and a detector followed by a low-pass filter and an amplifier. The input and output amplifiers can be set with a resolution of 2 dB.

The modulated signal to be analysed is, after proper amplification, mixed with a local carrier in a multiplier. The local carrier frequency is equal to the sum of the carrier frequency of the signal and the centre frequency of the tuned filter. The tuned filter will allow only the lower side band. The signal from a tuned filter is passed through a detector to recover the envelope of the signal. A low-pass filter further filters this. The envelope detector and the system are schematically shown in Fig. 1.3(a) and (b), respectively.
Brief technical particulars of the detector are as follows:

- Input signal: 10 mV rms (minimum)
- Carrier frequency of the signal: 20 kHz (maximum)

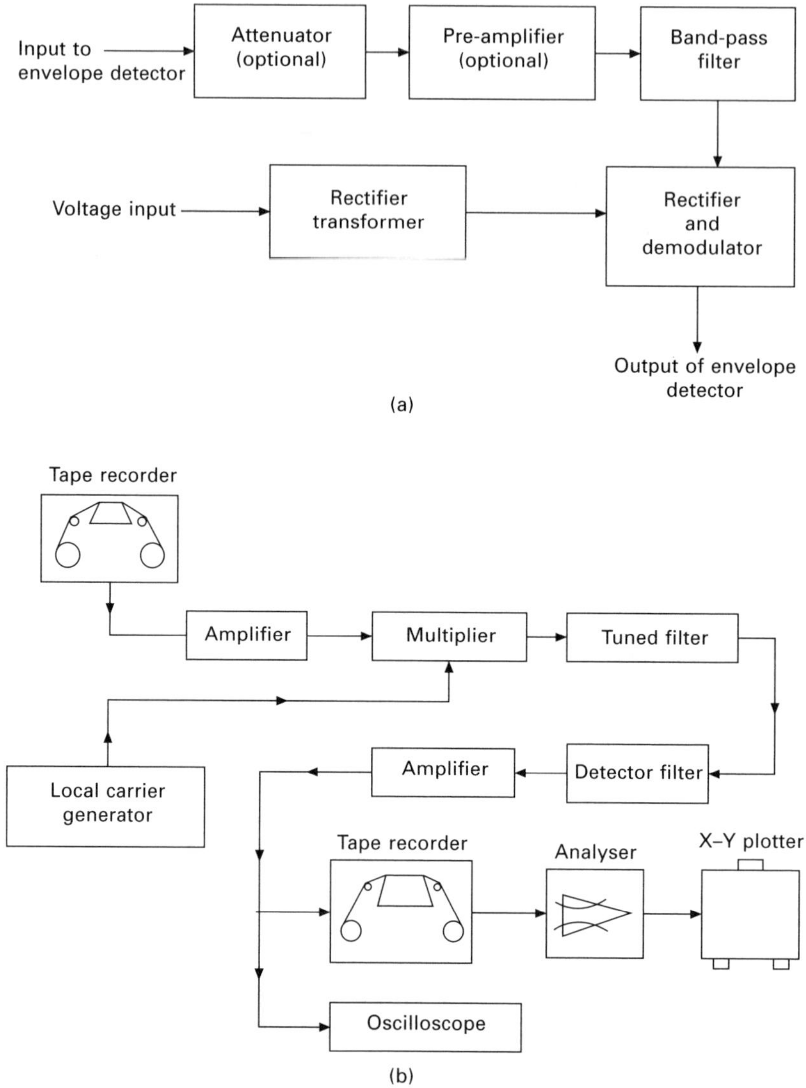

1.3 (a) Schematic of envelope detector; (b) envelope detection system.

- Bandwidth: 350 Hz
- Dynamic range: 40 dB

The bandwidth of the band pass filter is kept at double the defect frequency of interest. Usually ± 5 per cent is sufficient to pass the necessary side bands of the carrier frequency. If the bandwidth is set too large, unnecessary noise and distortion are liable to be passed by the filter, along with the carrier and side bands. The filtered signal as such is rectified and demodulated, to extract the envelope of the modulated carrier frequency signal. The amplitude of defect frequency of interest, as it appears in the frequency spectrum, can then be used to indicate directly the condition of the rolling-element bearing being tested.

In short, the vibration signature data recorded on the tape was demodulated using HFRT and re-recorded on another magnetic tape recorder. These re-recorded vibration signature data were further analysed by using a real-time analyser, and plots were taken on the plotter for further evaluation of the bearing defect frequencies.

1.7 Determination of defect frequencies and energy levels

1.7.1 Theoretical

Defect frequencies at different speeds of operation were theoretically determined for the NU 330 type of bearing by using the formulae given in Section 1.3. Four bearings of type NU 330, under identical conditions of operation in the coupled mode for the motor, were considered: bearing 'A' (new), bearing 'A' (defective), bearing 'B' (new) and bearing 'C' (defective). The defects of bearings 'A' (defective) and 'C' (defective) were known, bearing 'A' (defective) having a through crack in the inner race and bearing 'C' (defective) having score marks on the rollers. Table 1.1 shows the defect frequencies at different speeds of operation for NU 330 bearings.

1.7.2 Using the high-frequency resonance technique

Based on the calculated resonant frequencies, as given in Section 1.4, and the predominant frequencies existing in the raw signature, the significant centre frequencies were established. Then detailed signature analysis of bearings was carried out by using HFRT, with the significant centre frequencies ranging from 0 to 10 kHz at various speeds of operation under coupled mode. The frequency analysis was made and the plots were drawn in the range 0 to 500 Hz. The number of averages in the analysis was taken as 64, as shown in Fig. 1.4 to 1.10. Energy levels were calculated for significant defect frequencies, based on the signal conditioner sensitivity ($mV\,s^2m^{-1}$ rms) and input demodulation level (dB) for the respective frequencies. Energy levels are arbitrarily used as the acceleration levels in ms^{-2} rms times the unit total amplitude (voltage) ratio of the demodulated signals comprising defect frequencies. This does not carry any reference to other standard systems of units. The predominant defects existing in each bearing along with their respective energy levels are shown in Tables 1.2 to 1.5 for bearings 'A' (new), 'A' (defective), 'B' (new) and 'C' (defective), respectively. Figures 1.11 to 1.13 show raw vibration signatures without demodulation.

Table 1.1 Theoretical determination of defect frequencies at different operating speeds for NU 330 bearing

Defect frequency	Frequency (Hz) at different speeds of operation (rpm)											
		1000			1800			2250			2750	
f_s		16.67			30.00			37.50			45.83	
f_c		6.67			12.0			15.2			18.3	
f_b		42.5			76.5			95.8			116.9	
f_{ORDFL} (L_1, L_2)		93.4			168.0			214.0			256.0	
		186.8			336.0			428.0			513.0	
f_{bf}		85.0			153.0			191.0			235.8	
f_{IRDFL} (L_1, L_2)		139.0			250.0			318.0			382.0	
		278.0			500.0			636.0			764.0	
f_{IRWL} (L_1, L_2)	156.0		295.0	280.0		530.0	356.0		674.0	428.0		810.0
	122.0		261.0	220.0		470.0	280.0		598.0	336.0		718.0
f_{IRWNL} (L_1, L_2)	16.67		26.67	30.0		48.0	38.0		61.0	46.0		73.0
	3.34		13.33	6.0		24.0	7.4		30.5	9.2		36.5
f_{ORDFNL} (L_1, L_2)	0.0		6.7	0.0		12.0	0.0		15.0	0.0		18.3
	13.34		20.0	24.0		36.0	30.5		46.0	36.6		37.0
f_{RW} (W_1, W_2)	91.7		176.7	165.0		318.0	210.0		405.0	242.0		486.0
	78.4		163.4	141.0		294.0	180.0		374.0	215.0		445.0
f_{RRS} (S_1, S_2)		85.0			153.0			195.0			243.0	
		170.0			306.0			390.0			466.0	

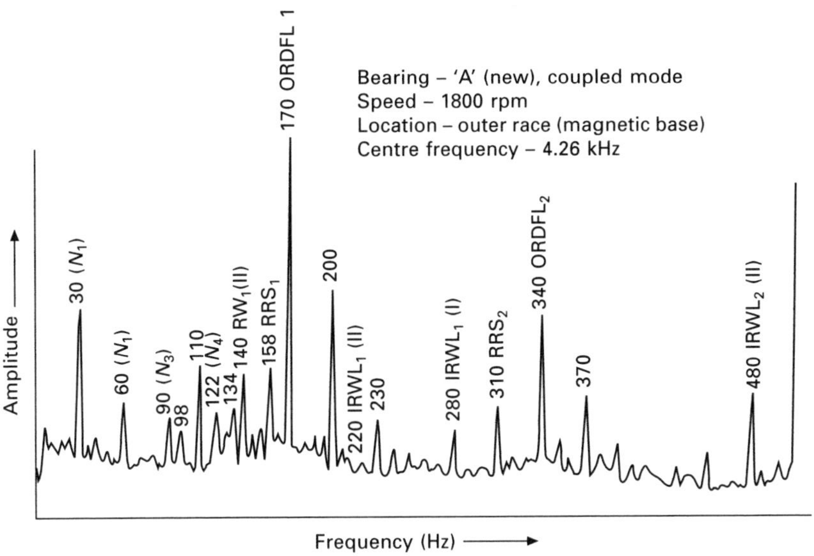

1.4 Frequency spectrum after demodulation for 165M traction motor.

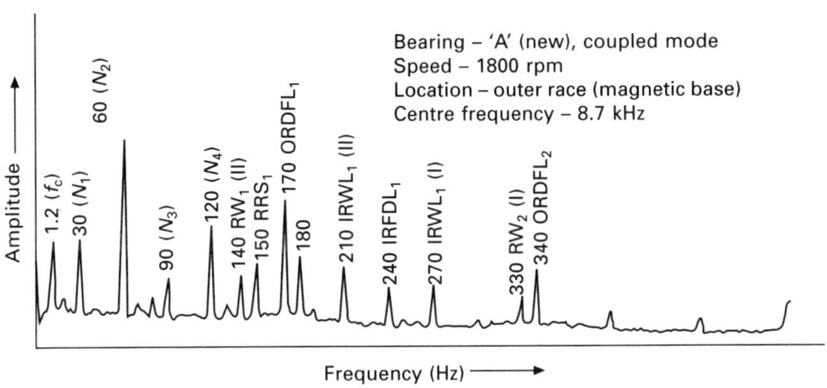

1.5 Frequency spectrum after demodulation for 165M traction motor.

1.8 Results and discussion

Tables 1.2 to 1.5 and Fig. 1.4 to 1.13 show the following:

- In the spectrum of bearing 'A' (new), certain defect frequencies such as RW, ORDFL, RRS, IRWL appear (Table 1.2), but these defects have significantly low energy levels compared with the energy levels exhibited by the same defects of bearing 'B' (new) (Table 1.4).
- Predominant defect frequencies appear in the coupled mode of operation under different speeds of operation.

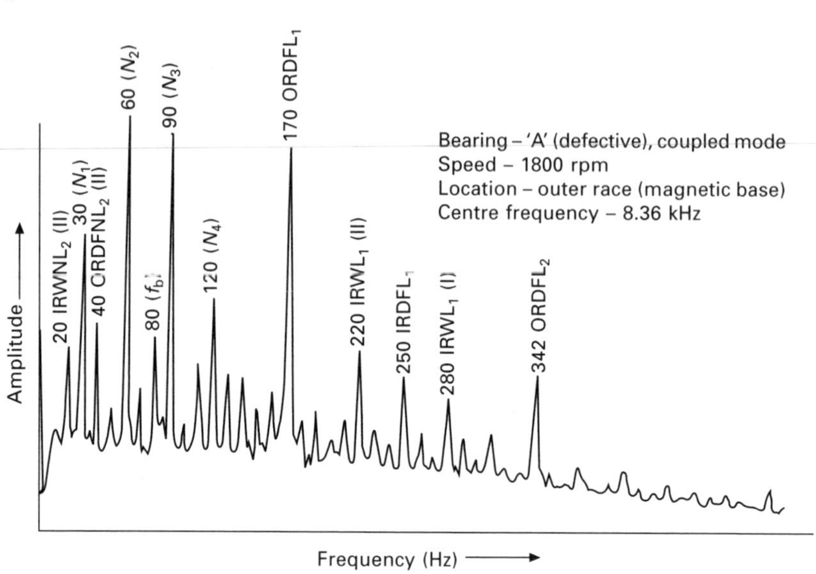

1.6 Frequency spectrum after demodulation for 165M traction motor.

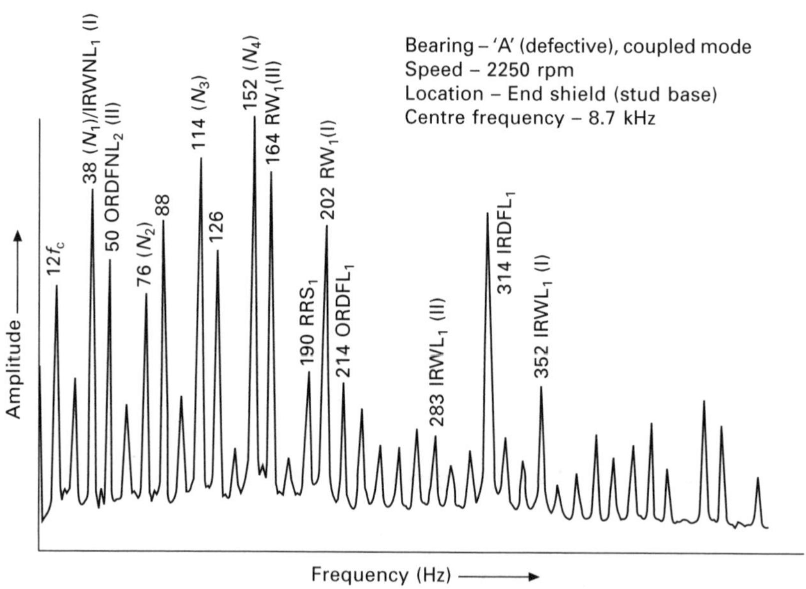

1.7 Frequency spectrum after demodulation for 165M traction motor.

- Energy levels of defect frequencies, in general, increase as the speed of operation increases to the maximum operating speed. For example, the energy level of 44.0 at 1000 rpm corresponding to the $ORDFL_1$ defect

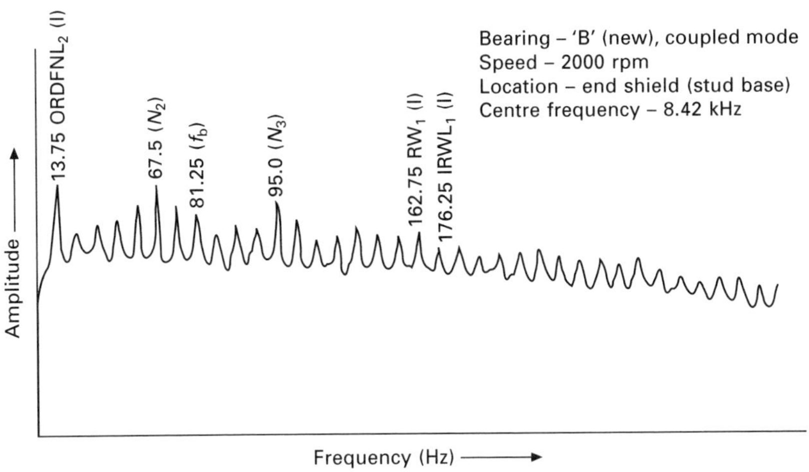

1.8 Frequency spectrum after demodulation for 165M traction motor.

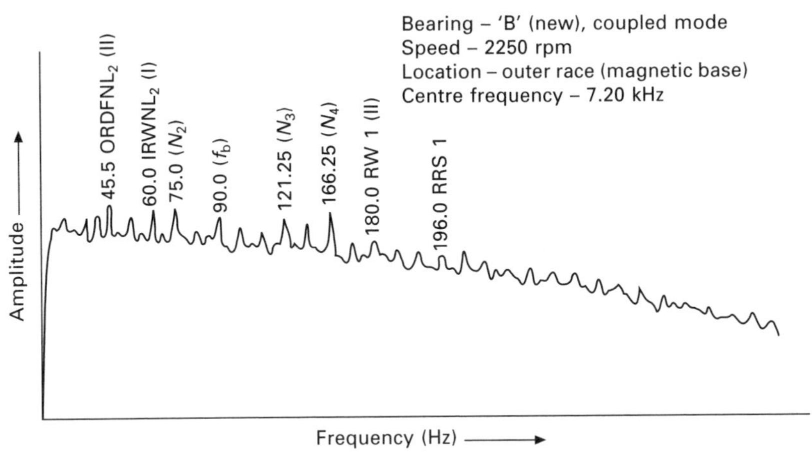

1.9 Frequency spectrum after demodulation for 165M traction motor.

(Table 1.2) increases from 80.9 at 1800 rpm to 296.5 at 2250 rpm from the magnetic base pick-up mounted on the bearing outer race for bearing 'A' (new). Similarly for bearing 'A' (defective) energy levels corresponding to the $IRWL_1$ (II) defect increases to 1246 at 2250 rpm from 74 at 1000 rpm (Table 1.3). A similar trend holds good for bearing 'B' (new).

- The change of centre frequency in HFRT analysis does not appreciably alter the nature of the frequency distribution in the spectrum.
- A raw spectrum without demodulation does not reveal the intrinsic condition of a bearing (Fig. 1.11 to 1.13).
- Stud base and magnetic base pick-ups mounted on the bearing outer race

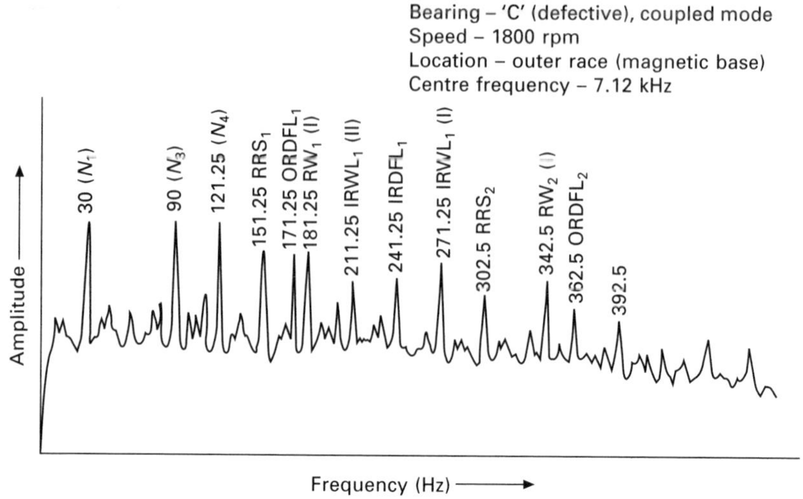

1.10 Frequency spectrum after demodulation for 165M traction motor.

show more or less identical behaviour so far as frequency distribution is concerned. A magnetic base pick-up directly mounted on the bearing outer race indicates somewhat higher energy levels for defect frequencies, because it avoids the effect of passage on the final signal received.

- Plots pertaining to defective bearings are more complex than those in respect of new bearings. This is due to the presence of different harmonics of defect frequencies, rotating speed frequencies and other complex response interactions.

1.8.1 Comparison of the performance of the bearings

The relative comparison of the bearings 'A' (new), 'A' (defective) and 'B' (new) at 2250 rpm of operation under coupled mode, in respect of the significant defects, is given in Table 1.6. From the values of energy levels of different defect frequencies, the performance of the bearing 'B' is found to be much inferior to that of bearing 'A' (new). From Table 1.6, it is evident that the performance of bearing 'A' (defective) is inferior to those of the other bearings tested. The energy levels were found to be very high (up to 1246), corresponding to defect frequencies for inner race, roller waviness, roller rough spots and outer race.

Comparison of the performance of bearing 'A' (new) with that of bearing 'C' (defective) at 1800 rpm of operation is given in Table 1.7. The presence of score marks on the rollers of bearing 'C' (defective) indicates severity of roller defects in terms of higher values of energy levels for roller waviness and roller rough spots (RW, RRS) in comparison to bearing 'A' (new). From

Table 1.2 Frequency analysis of 'A' (new) bearing NU 330 of 165M traction motor under coupled mode of operation

Speed (rpm)	1000			1800					2250			Predominant defect matching frequency	Remarks
Pick-up (location)	Magnetic base (outer race)			Magnetic base (outer race)					Magnetic base (outer race)				
CF (kHz)	3.46			4.26		8.7			4.52				
	Frequency (Hz)	EL	DMF	Frequency (Hz)	EL	Frequency (Hz)	EL	DMF	Frequency (Hz)	EL	DMF		
	16	74.26	N_1	30	111.5	30	60.0	N_1	38	188	N_1	ORDFL at 1000, 1800 and 2250 rpm	Outer race defects in microscopic nature
	30	40.30		60	61.8	60	112.5	N_2	78	179	N_2		
	34	43.3	N_2	90	53.7	90	36.9	N_3	116	179	N_3		
	38	28.6		98	49.8								
	54	32.2	N_3	110	80.1				154	169.5	N_4		
	68	35.9	N_4	122	57.0	120	67.5	N_4	214	296.5	$ORDFL_1$		
	72	31.1	RW_1(II)	134	58.7								
	96	44.0	$ORDFL_1$	140	77.5	140	40.9	RW_1 (II)					
	102	28.3		158	79.8	150	45.3	RRS_1					
				170	200.0	170	80.9	$ORDFL_1$					
				200	120.0	180	50.2						
				220	33.2	210	45.9	$IRWL_1$((II)					
				230	53.5								
						240	35.4	$IRDFL_1$					
				280	47.7	270	35.7	$IRWL_1$(I)					
				310	58.5			RRS_2					
						330	28.7	RW_2(I)					
				340	108.5	340	42.8	$ORDFL_2$					
				370	66.2								
				480	66.7			$IRWL_2$(II)					
						12	58.3	f_c					
				(See Fig. 1.4)		(See Fig. 1.5)							

CF, centre frequency; DMF, defect matching frequency; EL, energy level

Table 1.3 Frequency analysis of 'A' (defective) bearing NU 330 of 165M traction motor under coupled mode of operation

Speed (rpm)	1000					1800					2250						
Pick-up (location)	Stud base (end shield)		Magnetic base (outer race)			Stud base (end shield)		Magnetic base (outer race)			Stud base (end shield)		Magnetic base (outer race)			Predominant defect matching frequencies	Remarks
CF (kHz)	8.34		6.98			8.5		8.36			8.7		6.66				
	Frequency (Hz)	EL	Frequency (Hz)	EL	DMF	Frequency (Hz)	EL	Frequency (Hz)	EL	DMF	Frequency (Hz)	EL	Frequency (Hz)	EL	DMF		
	16	14.6	16	82.0	N_1	10	49.6			f_c	12	236.5			f_c		Inner
	34	18.5	34	117.0	N_2			20	62.0	$IRWNL_2$(II)	26	155.0					race
	50	15.7	50	98.0	N_3	30	80.26	30	99.3	N_1	38	321.5	38	966	$IRWNL_1$(I)		defects
	56	11.9	56	80.6				40	60.4	$ORDFNL_2$(II)	50	257	50	848	$ORDFNL_2$(II)		and
			72	69.2	RW_1(II)	60	48.8	60	139.4	N_2	64	135.6					roller
	84	12.5			RRS_1			80	66.7	f_b	76	226.5	76	938	N_2		defects
	90	12.0	90	88.5	RW_1(I)	90	40.3	90	132.0	N_3	88	290	88	1000			
			106	71.4		110	45.2	110	58.0		114	348	114	1634	N_3		CF - Centre
			124	74.0	$IRWL_1$(II)	120	73.6	120	78.9	N_4	126	265	126	1106			frequency
			140	75.0	$IRDFL_1$	130	41.7				152	383.3	152	614	N_4		EL - Energy
						140	40.9			RW_1(II)	164	332.5	164	1098	RW_1(II)		level
						150	49.3			RRS_1	190	160.0	190	788	RRS_1		DMF -
						160	51.2			RW_1(II)	202	286.5	202	1176	RW_1(I)		Detect
						170	44.2	170	128.4	$ORDFL_1$	214	147.5			$ORDFL_1$		matching
						222	29.0	220	63.0	$IRWL_1$(II)	283		283	1246	$IRWL_1$(II)		frequency
						250	44.9	250	53.6	$IRDFL_1$	314	297.5	314		$IRDFL_1$		
						280	29.2	280	46.05		352	148.0			$IRWL_1$(I)		
						342	28.44	342	54.0	$ORDFL_2$	466	133.5					

Table 1.4 Frequency analysis of 'B' (new) bearing NU 330 of 165M traction motor under coupled mode of operation

Speed (rpm)	2000			2250			2750			Predominant defect matching frequencies	Remarks
Pick-up (location)	Stud base (end shield)			Magnetic base (outer race)			Magnetic base (outer race)				
CF (kHz)	8.42			7.20			7.08				
	Frequency (Hz)	EL	DMF	Frequency (Hz)	EL	DMF	Frequency (Hz)	EL	DMF		
	13.72	471.0	$ORDFNL_2$(I)	45.0	636.0	$ORDFNL_2$(II)	12.5	262.0	$IRWNL_1$(II)	ORDFNL, IRWNL, IRWL, RRS and RW	Roller, outer and inner race defects
	67.5	305.0	N_2	60.0	531.0	$IRWNL_2$(I)	20.0	441.0	$ORDFNL_2$(I)		
	81.25	430.0	f_b	75.0	583.0	N_2	37.50	274.0	$IRWNL_2$(I)		
	95.00	334.0	N_3	90.0	608.0	f_b	63.75	274.0	$IRWNL_2$(I)		CF - Centre frequency
	108.75	293.0		121.25	512.0	N_3	161.25	381.5			
	162.75	307.0	RW_1(I)	166.25	531.0	N_4	221.25	252.0	RW_1(II)		EL - Energy level
	176.25	284.0	$IRWL_1$(I)	180.00	448.0	RW_1(II)					DMF - Detect matching frequency
				196.00	604.0	RRS_1					
	(See Fig. 1.8)			(See Fig. 1.9)							

Table 1.5 Frequency analysis of 'C' (defective) bearing NU 330 of 165M traction motor under coupled mode of operation

Speed (rpm)	1800					Predominant defect matching frequencies	Remarks
Pick-up (location)	Magnetic base (outer race)						
CF (kHz)	5.14		7.12				
	Frequency (Hz)	EL	Frequency (Hz)	EL	DMF		
	30	271.83	30	271.83	N_1	RW, ORDFL, RRS,	Roller waviness,
	91.25	250.34	90	258.83	N_3	IRWL and IRDFL	inner and outer
	121.25	310.84	121.25	271.83	N_4		race defects
	151.25	188.25	151.25	195.00	RRS_1		
			171.25	168.75	$ORDFL_1$		
	181.25	231.24	181.25	193.00	RW_1(I)		
			211.25	119.4	$IRWL_1$(II)		
	242.50	139.0	241.25	118.21	$IRDFL_1$		
	272.5	167.96	271.25	151.24	$IRWL_1$(I)		CF - Centre frequency
	302.5	116.22	302.5	95.12	RRS_2		EL - Energy level
	342.5	144.47	342.5	122.19	RW_2(I)		DMF - Defect matching
	362.5	91.54	362.5	84.38	$ORDFL_2$		frequency
	392.5	91.54	392.5	84.38			
	423.75	89.55			$IRWL_1$(II)		
		(See Fig. 1.10)					

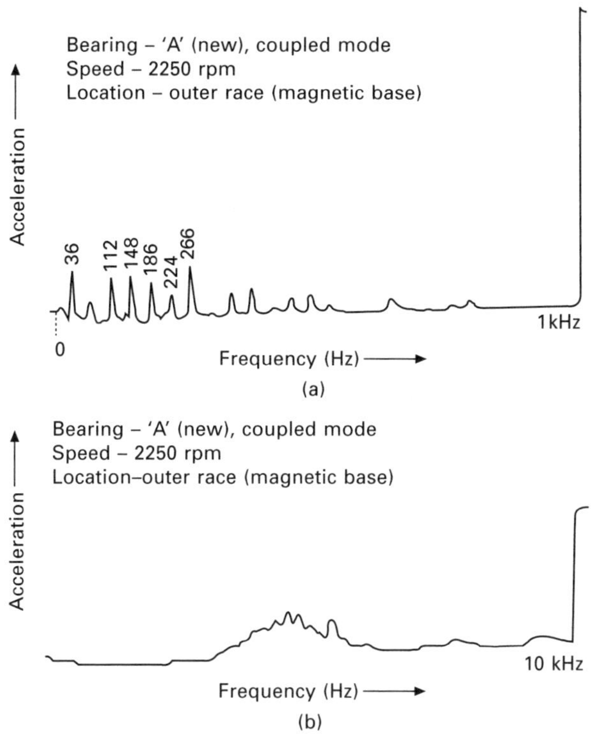

1.11 Raw signature without demodulation. Frequency range (a) 0–1 kHz; (b) 0–10 kHz.

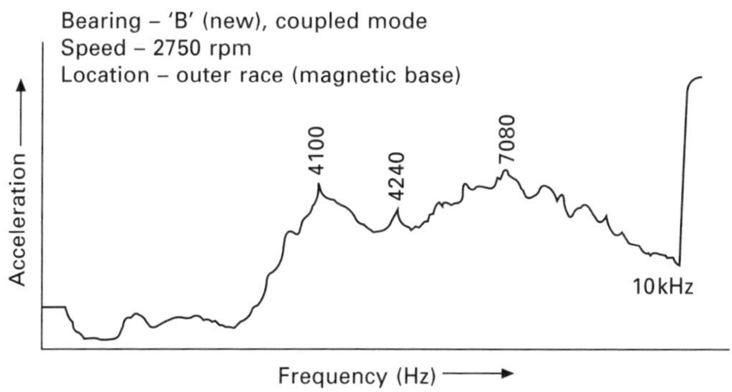

1.12 Raw signature without demodulation.

Table 1.7, it is evident that the energy levels of defect frequencies of bearing 'C' (defective) are two to four times higher than those of bearing 'A' (new). Bearing 'C' (defective) has significant defects in rollers as well as in inner and outer races.

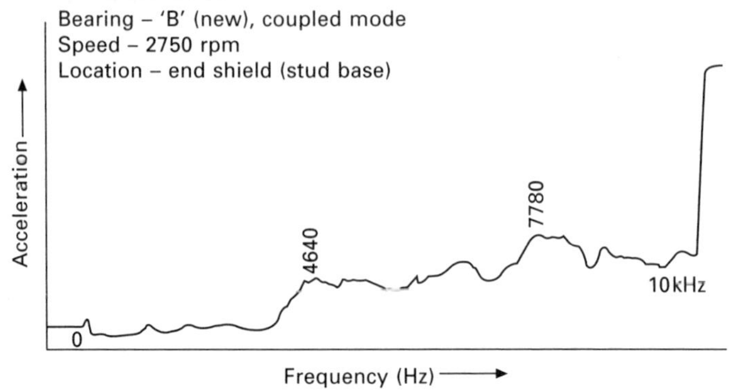

1.13 Raw signature without demodulation.

Table 1.6 Comparison of the performance of bearings 'A' (new), 'A' (defective) and 'B' (new) at 2250 rpm (data from magnetic base pick-up mounted on bearing outer race)

Defect	Energy level		
	'A' (new) (4.52 kHz CF)	'B' (new) (7.20 kHz CF)	'A' (defective) (6.66 kHz CF)
$ORDFL_1$	296.5	–	–
$ORDFNL_2$(II)	–	636.0	848.0
$IRWNL_2$(I)	–	531.0	–
$IRWNL_1$(I)	–	–	966.0
$IRWL_1$(II)	–	–	1246.0
RW_1(II)	–	448.0	1098.0
RW_1(I)	–	–	1176.0
RRS_1	–	604.0	788.0

Table 1.7 Comparison of the performance of bearings 'A' (new) and 'C' (defective) at 1800 rpm (data from magnetic base pick-up mounted on bearing outer race)

Defect	Energy level	
	'A' (new) (8.7 kHz CF)	'C' (defective) (7.12 kHz CF)
RW_1(II)	40.9	–
RW_2(I)	28.7	122.8
RRS_1	45.3	195.0
RW_1(I)	–	193.0
RRS_2	–	95.1
$ORDFL_1$	80.9	168.8
$IRWL_1$(II)	45.9	119.4
$IRDFL_1$	35.4	118.2
$IRWL_1$(I)	35.7	151.2
$ORDFL_2$	42.8	84.4

An examination of experimental data (Tables 1.6 and 1.7) indicates that bearing 'A' (defective) has four to eight times higher energy levels for identified defect frequencies than bearing 'C' (defective).

1.9 Conclusions and recommendations

Based on this study and analysis, the following major conclusions are drawn:[8,9]

- Data brought out in this study can be used for appropriate bearing selection, performance evaluation and other relative comparisons for NU 330 bearings.
- From the results of various bearings by high-frequency resonance technique (HFRT), it is recommended that bearing 'B' (new) is not used since the performance of this bearing is much inferior to that of bearing 'A' (new). This is also confirmed by the field data, which indicate that the failure rate is as high as 25% for the bearing 'B' (new) as against 0.9% for the bearing 'A' (new).
- Energy levels of defect frequencies detected by HFRT for bearing 'A' (new) at the maximum speed of operation are, in general, less than 300. This may be taken as guide to study for establishing the severity of defects for bearings under test and also as the limit of trouble-free normal and healthy bearing operation.
- Continuous indication of energy levels of more than 300 for any defect frequency should be taken as serious and the concerned bearing may be treated as defective with incipient damage.
- Any bearing showing energy levels of more than 500 for any defect frequency should not be used to avoid catastrophic bearing seizure/failure.
- Vibration pick-up mounted on the bearing outer race avoids the effect of passage on the finally received signal. It is the best indication of bearing health, compared with the other pick-ups.
- To avoid a catastrophic bearing seizure, bearing 'B' (new) and bearing 'A' (defective) are not recommended for use, since energy levels of the detected defect frequencies are higher than 500 in the operating speed range.
- HFRT is recommended to be used for quality control of bearings on a test stand to identify even the smallest bearing defect, which cannot be identified by bearing temperature rise, overall vibration levels and shock pulse meter records.
- With the help of HFRT, a complete check on the quality of the bearing can be kept, and the specified life of the bearing under actual operation can be ensured.

1.10 References

1 Darlow, M.S. and Badgley, R.H., 'Applications for Early Detection of Rolling-Element Bearing Failures Using High Frequency Resonance Techniques', ASME, 75-DET-46, 1975.

2 Weichbrodt, B. and Smith, K.A., 'Signature Analysis Non-Intrusive Techniques for Incipient Failure Identification – Application to Bearings and Gears', *Proc. 5th Space Simulating Conf.*, 1970, National Bureau of Standards, Gaithersburg, MD.

3 Burchill, R.F., 'Resonant Structure Techniques for Bearing Fault Analysis', *Proc. 18th Meeting MFPG*, 1964, National Bureau of Standards, Gaithersburg, MD.

4 Darlow, M.S. and Badgley, R.H., 'Early Detection of Defects in Roller Element Bearings', SAE Paper 750209, 1975.

5 Babkin, A.S. and Anderson, J.J., 'Mechanical Signature Analysis of Ball Bearings', Application Note 3, December 1972, Federal Scientific.

6 Sramek, B., 'Frequency calculations for Ball Bearings', Preliminary note, April, 1978, Nicolet Scientific Corp.

7 Martin, R.L., 'Detection of Ball Bearing Malfunctions', *J. Instrum. Control Systems*, 79–82, Dec. 1970.

8 Prashad, H., *et al.*, 'Reliability Analysis of Anti-friction Bearings by High Frequency Resonance Technique', *BHEL J.*, 1985, **6**(1), 1–15.

9 Prashad, H., Ghosh, M. and Biswas, S., 'Diagnostic Monitoring of Rolling-Element Bearings by High Frequency Resonance Technique', *ASLE Trans.*, 1985, **28**(4), 439–448.

1.11 Nomenclature

α	contact angle
ρ	density of rolling-element
a	radius of neutral axis
D	pitch diameter
d	diameter of rolling-element
E	Young's modulus of elasticity
f_s	shaft rotational frequency
f_c	cage rotational frequency
f_b	rolling-element frequency
f_{br}	free resonance frequency of rolling-element
f_{bf}	rolling-element defect frequency
f_{rr}	resonance frequency of race
f_{ORDFL}	outer race defect frequency (linear)
f_{IRDFL}	inner race defect frequency (linear)
f_{IRWL}	inner race waviness frequency (linear)
f_{IRWNL}	inner race waviness frequency (non-linear)
f_{ORDFNL}	outer race defect frequency (non-linear)
f_{RW}	roller waviness frequency
f_{RRS}	roller rough spot frequency
f_{ORDFL} (L_1, L_2)	outer race defect frequency (linear) for $n = 1$ and $n = 2$

f_{IRDFL} (L_1, L_2)	inner race defect frequency (linear) for $n = 1$ and $n = 2$
f_{IRWL} (L_1, L_2)	inner race waviness frequency (linear) for $n = 1$ and $n = 2$
f_{IRWNL} (L_1, L_2)	inner race waviness frequency (non-linear) for $n = 1$ and $n = 2$
f_{ORDFNL} (L_1, L_2)	outer race defect frequency (non-linear) for $n = 1$ and $n = 2$
f_{RRS} (S_1, S_2)	roller rough spot frequency for $n = 1$ and $n = 2$
f_{RW} (W_1,W_2)	roller waviness frequency for $n = 1$ and $n = 2$
I	moment of inertia of cross-section
K	order of resonance
m	mass of race per unit linear length
n	1, 2, 3, …, 9 (harmonics)
N	number of rolling-elements
r	radius of rolling-element

Above frequencies with L_1(I), S_1(I), W_1(I) correspond to $n = 1$ for first order and L_1(II), S_1(II), W_1(II) to $n = 1$ for second order. Similarly, frequencies with L_2(I), S_2(I), W_2(I) correspond to $n = 2$ for first order and L_2(II), S_2(II), W_2(II) to $n = 2$ for second order.

2

Functional performance of rolling-element bearings for acceptance in routine applications

2.1 A general review

This chapter deals with the functional performance of roller bearings on a roller-bearing test rig for the acceptance of bearings in routine application in different products. The salient features of bearing test philosophy and analysis of test data collected on the tested bearings pertaining to temperature rise, vibration, shock pulse and kurtosis analysis under identical operating parameters in different stages of operation are bought out. The investigations reveal that under light and medium operating parameters the performance of identical bearings of different makes are in general within acceptable limits. However, under higher stipulated operating parameters, the behaviour and functional performance of identical bearings of different makes differ from each other, both in life expectancy and general performance characteristics.

2.2 Introduction

Demands on the reliable performance of various components, in particular rolling-element bearings, have increased considerably. High reliability and successful functioning of the bearings are prerequisites if complex machines and equipments are to operate satisfactorily. That is why newly designed and developed bearings are subjected to extensive testing to ensure maximum operational reliability in service.[1]

When planning reliability tests, the minimum laboratory test duration has to simulate as far as possible the actual service conditions. It is unwise to shorten the test duration by using an unrealistically high load, as this can incur the risk of the results being false. Also, under increased load, the shaft deflection will be larger, which will result in additional forces acting on the bearing, and may cause premature bearing failure.

The objective of the bearing test is therefore to carry out tailor-made functional tests under realistic load spectra in the shortest possible time in parallel with long-term endurance tests. Obviously, a laboratory rig test cannot completely replace the final acceptance test because it is not always possible to reproduce the effects of other components on the bearing performance.

Keeping in view the above aspects, the test programme for bearings with a limited test duration has been designed to be in line with those of the leading bearing manufacturers to evaluate the performance of various makes of bearings compared with the corresponding bearings of proven reliability. This study reports the periodical diagnostic test data, analysis and relative comparison of different makes of bearings.

2.3 Bearing test philosophy

In accordance with the present state of the art, bearing behaviour and performance analysis have been predominantly based on vibration, shock pulse value, kurtosis parameter and temperature rise. By monitoring these parameters at definite time intervals, the test bearing condition is monitored. Also, conclusions are drawn regarding comparative behaviour of rolling-element bearings of different makes operating under identical conditions.

2.3.1 Temperature

The bearing temperature is essential for the calculation of elastohydrodynamic film thickness, which is a critical parameter for determining the bearing fatigue life, assessment of moment of friction and generation of heat by a bearing. In general, heat generated by a bearing depends on load, speed, lubricant viscosity and bearing design, i.e. pitch diameter, diameter and length of rollers, and circular path configuration in roller tracks. These factors are considered as the measure of friction loss in a bearing.

In short, energy loss (P) in a bearing is proportional to the friction torque (M) and angular velocity (W), which can be expressed as:[2]

$$P = MW \tag{2.1}$$

Energy loss in a bearing gives rise to an increase in bearing temperature (ΔT) above ambient, which is determined as:

$$\Delta T = KP \tag{2.2}$$

or

$$\Delta T = KMW \tag{2.3}$$

where K is a constant depending on bearing speed and design factors. From the above, it is evident that bearing friction torque governs generation of heat and in turn influences the bearing performance.[3,4]

2.3.2 Vibration

Rolling-element bearings, while in operation, are subjected to damage that leads to a reduction in material strength or to a deterioration of the running qualities of a bearing. Thus, the reliability of a bearing is reduced, and a breakdown sooner or later is inevitable. Vibration signature analysis, crest factor analysis, power spectral density and overall vibration level techniques are used to determine the state of rolling-element bearings without dismantling, and to assess their condition in accordance with the requirements of the operating conditions.[5,7]

The high-frequency resonance technique (HFRT), a highly sensitive method for easy identification of an impending bearing failure, can be used to detect the location of even the smallest spall in a bearing (inner or outer race, or roller).[8,9]

2.3.3 Shock pulse meter

The shock pulse meter indirectly measures the velocity of mechanical impact, caused by a bearing irregularity, and allows a quantitative measurement of bearing condition without being influenced by vibration, noise, temperature or other external factors. The impact produces high-frequency sound vibrations in the range 30 to 50 kHz at the time of energy exchange between contact faces, i.e. roller/race. A slight surface damage brings about a large deviation in shock pulse levels. The instrument has been used to monitor the condition of bearings.[10]

2.3.4 Kurtosis

The kurtosis meter is used for bearing damage detection by measuring vibration levels using the kurtosis method for signal conditioning. It is independent of shaft speed, bearing load and size. The kurtosis value of normal Gauss distribution is 3. This is the distribution, that has been proved to relate to the acceleration vibration generated by a rolling-element bearing with no damage. The result of damage is to change the shape of the distribution by making it more peaky.

The peaky distribution produces a higher kurtosis value and increases with spread of damage. Using the sensitivity of five frequency bands (K_1 to K_5), varying between 2.5 to 80 kHz, the kurtosis factor K_A rises as damage increases.[11] Similar to kurtosis bands K_1 to K_5, acceleration levels g in rms (G_1 to G_5) are monitored under different stages of operation. G, in rms on a trend basis, is used to support the K_A factor and vibration velocity, V. In general, g indicates the time history of vibration acceleration levels.[11]

2.4 Bearing test machine

The bearing test machine shown in Fig. 2.1 and 2.2 was used to evaluate the performance of rolling-element bearings under different combinations of radial and axial loads. Bearings of different sizes (of inner diameter ranging from 44 to 150 mm) can be tested, and investigations in various tribological

2.1 Bearing test rig with control panel.

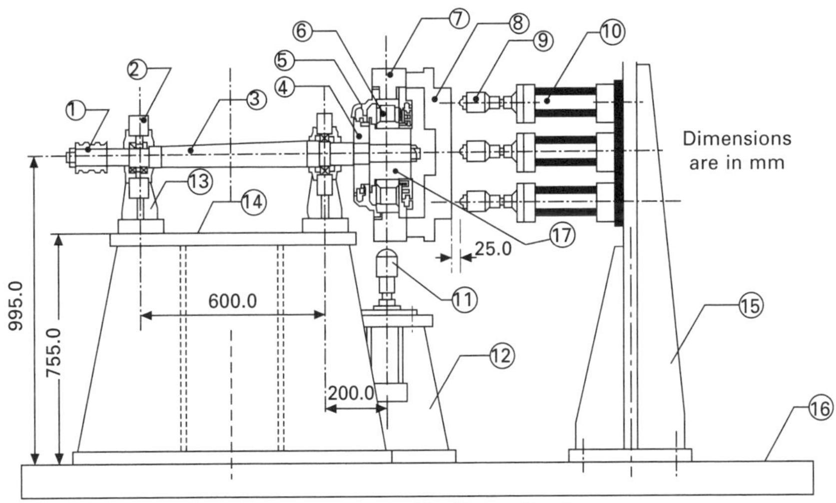

1. Drive pulley
2. Support bearing housing
3. Test shaft
4. Test bearing inner wiper
5. Test bearing inner cap
6. Test bearing
7. Test bearing housing
8. Axial load transmitting frame
9. Load cell
10. Hydraulic piston
11. Load cell
12. Support frame for hydraulic piston (radial)
13. Support bearing
14. Base pedestal for support bearings
15. Support frame for hydraulic loading (axial)
16. Bed plate
17. Test bearing sleeve

2.2 Sectional arrangement of roller-bearing test rig.

areas programmed. The condition of a test bearing is monitored by temperature rise, vibration analysis, and shock pulse and kurtosis data. The housing of the test bearing is specially designed to monitor signals directly from the outer race. The temperature scanner periodically, as per the set interval, automatically monitors grease temperature and outer race temperature of the test bearing.

To simulate the stipulated high operating parameters, flexible heaters, installed on the bearing housing, heat the test bearing. The desired temperature of the outer race is maintained by an automatic supply control to the heaters by the Servotron controller. As the bearing's outer race temperature is increased beyond the set value, the supply to the heaters is automatically cut off. Thus, the test bearings can be heated to different set temperatures to achieve the various operating parameters for different time intervals.

2.5 Experimental procedure

In the present series of investigations, a roller bearing of make 'A' and a similar bearing of make 'B' type 130-09-31180 were tested under identical conditions, and the condition of the bearings has been periodically monitored. The bearings have been lubricated with Servogem 3 grease. The internal dimensions of both 'A' and 'B' bearings are given in Table 2.1.

The bearings have been tested separately at 1000 rpm under 10000 and 5000N radial loads (acting at two positions 90° to each other) for 150h in five stages as per the following sequence:

1. up to 50 h under the above operating parameters;
2. from 50 to 75h by maintaining the bearing outer race temperature at 70 °C without changing the above operating parameters;
3. from 75 to 90h by maintaining the bearing temperature at 72°C;

Table 2.1 Internal dimensions of 'A' and 'B' bearings Type 130-09-31180

1	Inner diameter of inner race	= 90 mm
2	Outer diameter of outer race	= 190 mm
3	Width of bearing	= 43 mm
4	Outer diameter of inner race	= 115 mm
5	Inner diameter of outer race	= 165 mm
6	Length of roller	= 25 mm
7	Diameter of roller	= 25 mm
8	Pitch diameter	= 140 mm
9	Internal clearance; C4, class 90–140 (in μm)	
10	No. of rollers in 'A' bearing	= 13
11	No. of rollers in 'B' bearing	= 12
12	Material of bearings as specified and overall manufacturing tolerances as measured are identical	

4. from 90 to 110h at the bearing temperature of 73 °C; and
5. from 110 to 150h by maintaining the bearing temperature at 75 °C. However, the grease temperature was continuously monitored so as not to exceed 95 °C.

The above sequence of testing is principally adopted based on the 'FAG' bearing test procedure under accelerated testing.[12] The constant outer race temperature of 70–75 °C gives simulated operating parameters of operation as a bearing load of approximately 1 to 2 MPa.

2.6 Data deduction

2.6.1 Temperature rise of bearings

The temperature rises (above ambient) of housing, outer race, and grease of 'A' and 'B' bearings have been worked out, and plotted against the run time of different stages of sequential testing. Figures 2.3 to 2.5 indicate maximum

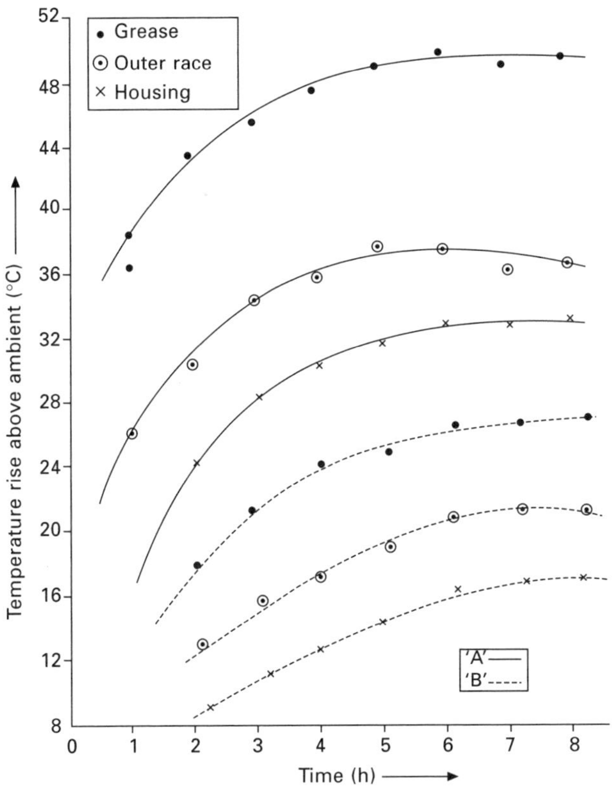

2.3 Maximum rise in temperature with time of housing, outer race and grease of 'A' and 'B' bearings (130-09-31180) up to 50h of operation at 1000 rpm under 10 000 N and 5 × 1000 N of radial loads.

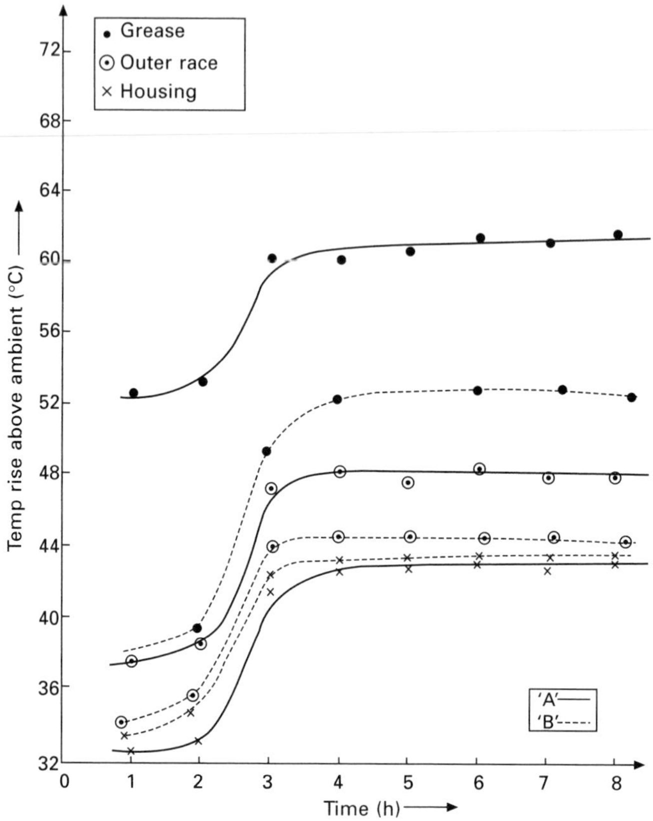

2.4 Maximum rise in temperature with time of housing, outer race and grease of 'A' and 'B' bearings (130-09-31180) in the range of 50 to 75 h at the maximum set temperature of outer race of 70 °C.

variation of temperatures of the bearings ('A' and 'B') with run time in stages 1 to 3, respectively. Figure 2.6 indicates the variation of temperature with time in stages 4 and 5, i.e. from 90 to 150 h of operation.

2.6.2 Shock pulse levels

The shock pulse levels (dbn) on outer race and housing with respect to run time of 'A' and 'B' bearing are shown in Fig. 2.7.

2.6.3 Kurtosis value

Variation of kurtosis values of bands K_1 to K_4, and G_2 to G_4 with run time of both makes of the identical bearings is shown in Figs. 2.8 and 2.10, respectively;

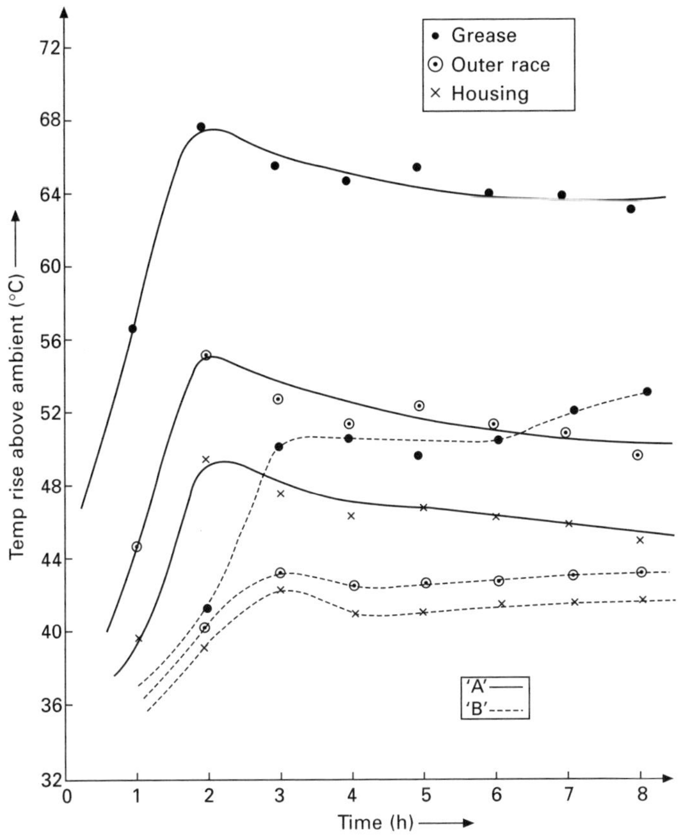

2.5 Maximum rise in temperature with time of housing, outer race and grease of 'A' and 'B' bearings (130-09-31180) in the range of 75 to 90 h at the maximum set temperature of outer race of 72 °C.

variation of kurtosis values K_5, K_A, V, and G_1 with run time are shown in Fig. 2.9 for 'A' and 'B' bearings.

2.6.4 Overall vibration levels and frequency spectrum

Figure 2.11 indicates the variation of overall vibration levels with run time at different stages for both makes of bearing. Figures 2.12 to 2.14 show frequency spectra at different run times.

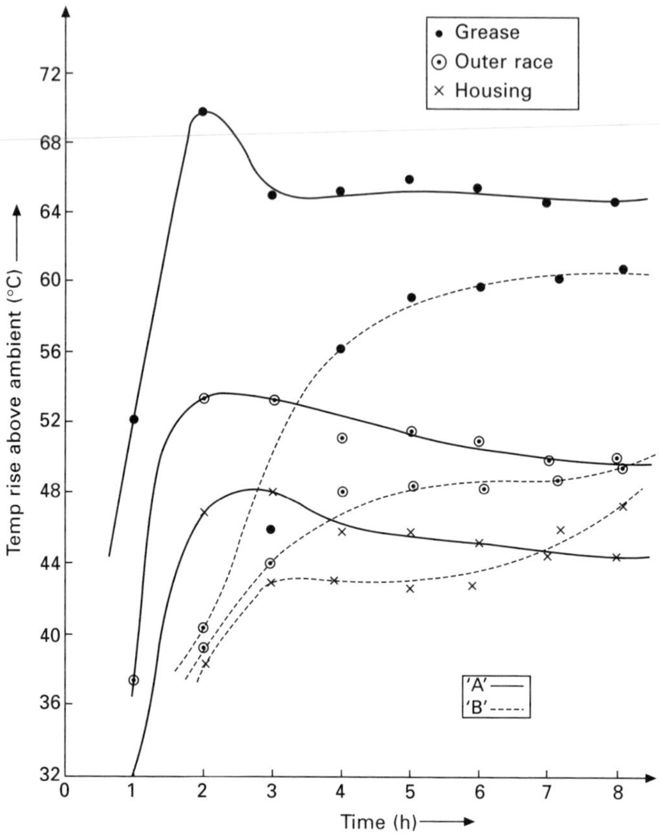

2.6 Maximum rise in temperature with time of housing, outer race and grease of 'A' and 'B' bearings (130-09-31180) in the range of 90 to 150 h at the maximum set temperature of outer race of 75 °C.

2.7 Results and discussion

2.7.1 Effect of rolling friction on temperature rise under different stages of operation

Up to 50 h of operation

It is evident from Fig. 2.3 that the maximum rises of housing, outer race and grease temperatures (above ambient) with time of the 'A' bearing up to 50 h of operation in stage 1 are stabilized at 32.4, 36.2 and 49.5 °C (above ambient) as against that of the 'B' bearing at 17.6, 21 and 26.3 °C, respectively. The higher rise of temperature of 'A' bearing may be due to predominant cage and roller slip because of a greater number of rolling-elements, causing less unit load compared with that of the 'B' bearing.[9]

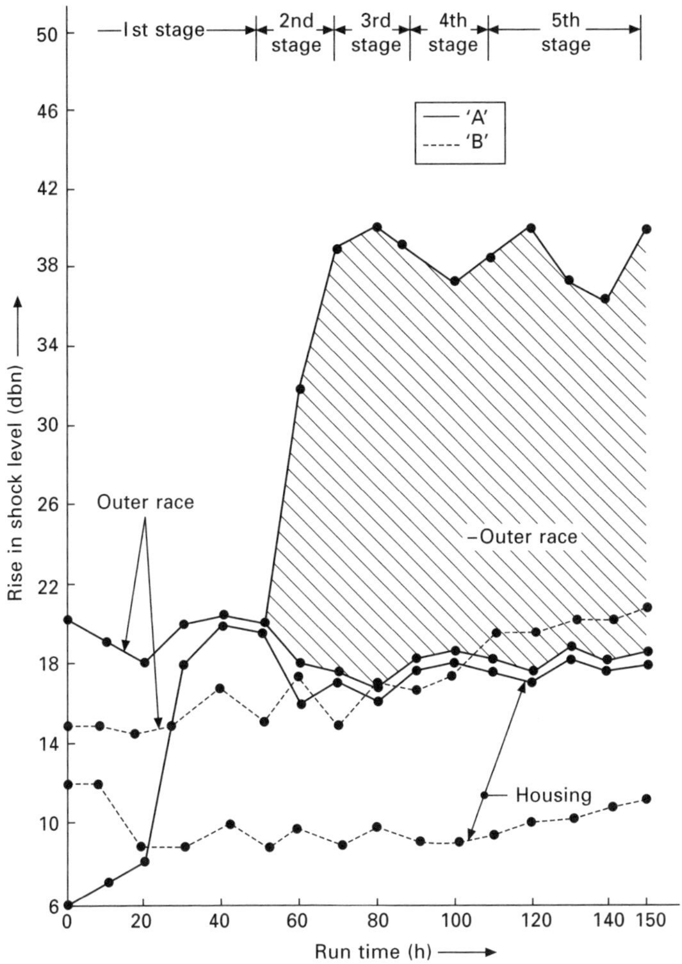

2.7 Variation of shock pulse (dbn) on outer race and housing versus run time of 'A' and 'B' bearings

Between 50 and 75 h of operation by maintaining outer race temperature at 70 °C

As above, the 'A' bearing housing, outer race and grease temperatures are stabilized at 43.5, 48 and 61.5 °C (above ambient) as against that of the 'B' bearing at 44, 44.5 and 52.5 °C, respectively, (Fig. 2.4) by external heating of the bearing housing at constant set temperature as discussed above.

Between 75 and 90 h of operation by maintaining the outer race temperature at 72 °C

Under the identical operating conditions by maintaining temperature at

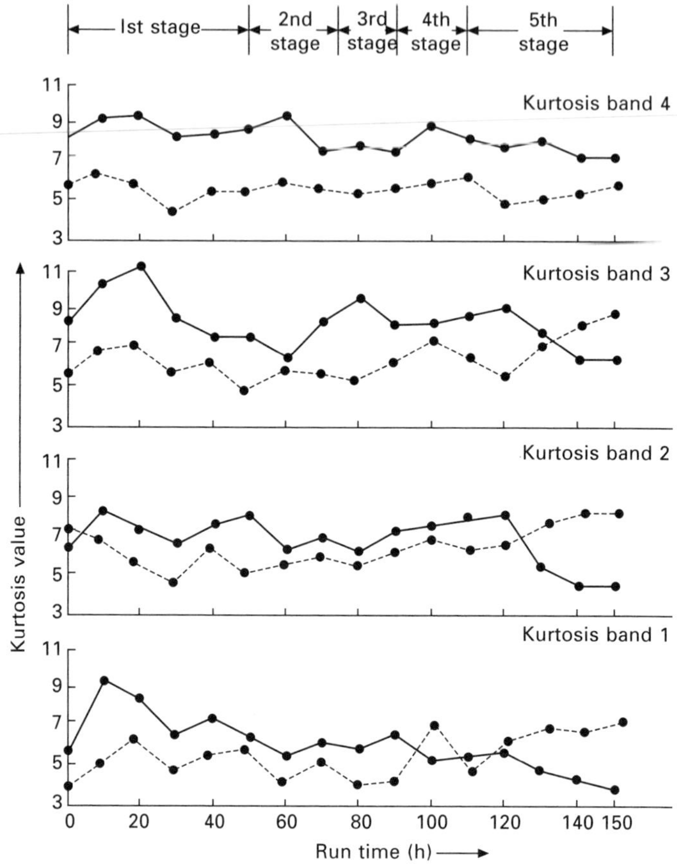

2.8 Variation of kurtosis values of bands K_1 to K_4 with run time for 'A' (—) and 'B' (- - -) bearings.

72 °C, the 'A' bearing housing, outer race and grease temperatures were stabilized at 45.5, 50 and 63.5 °C (above ambient) respectively, as shown in Fig. 2.5. In contrast, 'B' bearing temperatures were lower, but these temperatures were not found to stabilize and had a rising trend during this period of operation. This indicates the inferior behaviour of the 'B' bearing at higher operating parameters.

Between 90 and 110h to 150h of operation by maintaining the outer race temperature at 75° C

Under similar operating conditions as above by maintaining the outer race temperature at 75 °C, the 'A' bearing housing, outer race and grease temperatures were stabilized at 45, 50 and 65 °C (above ambient), respectively, (Fig. 2.6).

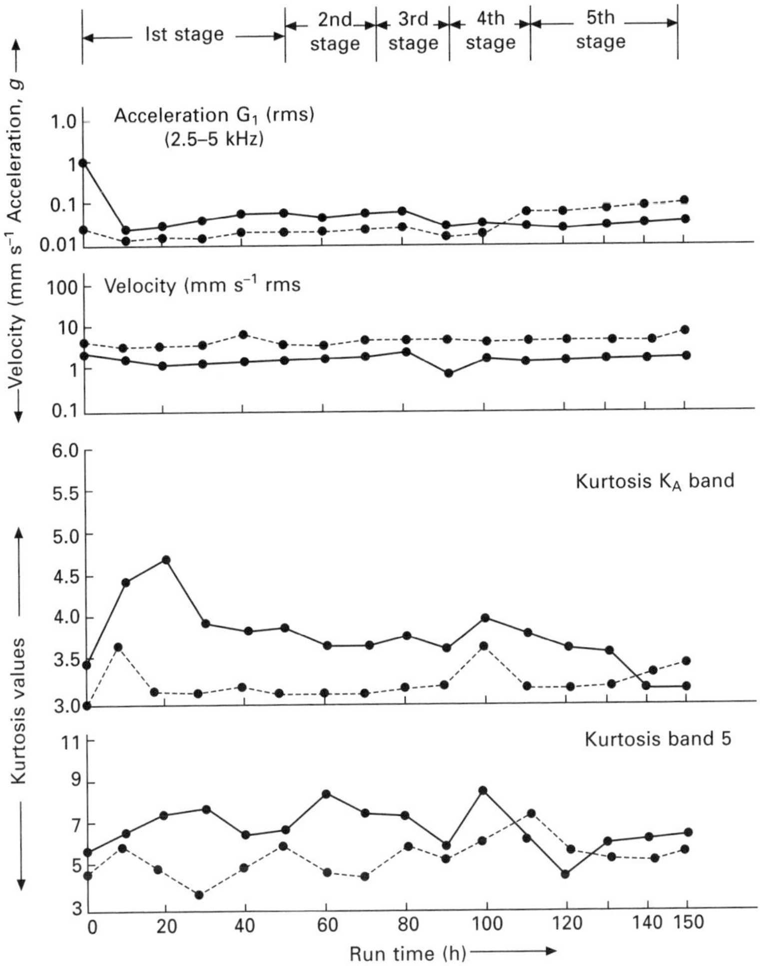

2.9 Variation of kurtosis values of K_5, K_A, V G_1 and with run time for 'A' (—) and 'B' (- - -) bearings.

'B' bearing temperatures were not stabilized, and housing, outer race and grease temperatures were recorded as 48, 52 and 61.5 °C, respectively. This further confirms the inferior quality of the 'B' bearing.

2.7.2 Shock pulse level of 'A' and 'B' bearings under different stages of operation

The dbn levels of the 'A' bearing detected from outer race and housing varied between 16 and 20 up to 150 h of operation (including that of maximum set temperature of 75 °C of outer race), which indicates that the bearing

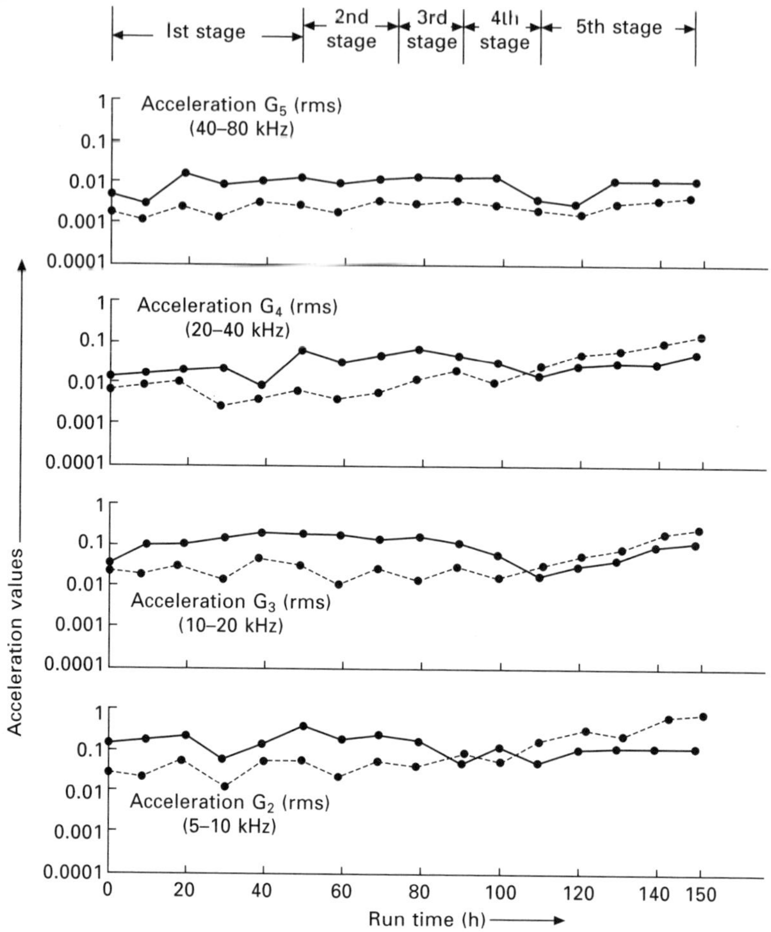

2.10 Variation of Kurtosis values of G_2, G_3, G_4 and G_5 with run time for 'A' (—) and 'B' (- - -) bearings.

remained in good condition throughout the operating period and no deterioration had taken place. However, during heating of the bearing, an intermittent parallel peak from the bearing outer race around 40 dbn was detected, which indicates a lack of lubrication in the bearing (Fig. 2.7); the bearing did not show any sign of deterioration/damage on visual inspection after the completion of test. The dbn levels of the 'B' bearing from the outer race and housing were less than 18 and 10 up to the third stage of operation. In the fourth and fifth stages of operation, the dbn levels increased beyond 21 and 13. This indicates the initiation of incipient bearing damage. On visual inspection after the test, the 'B' bearing showed score marks on the rolling surfaces.

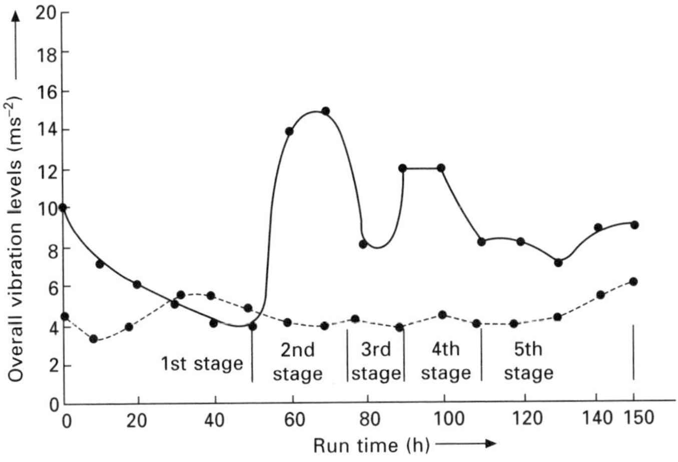

2.11 Variation of overall vibration levels (ms^{-2}) with time for 'A' (—) and 'B' (- - -) bearings.

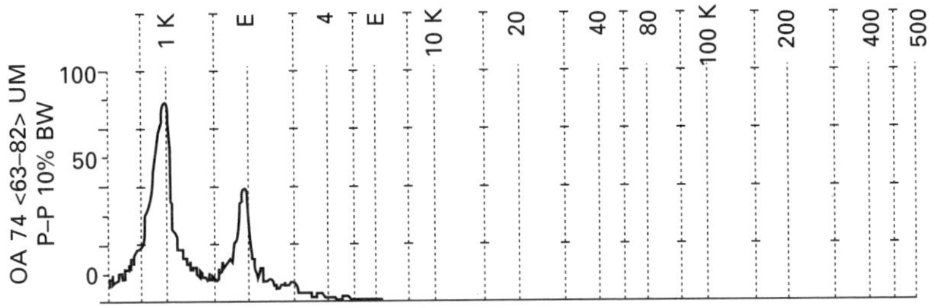

2.12 Frequency spectrum of 'A' bearing (outer race) in amplitude mode after 86 h of operation.

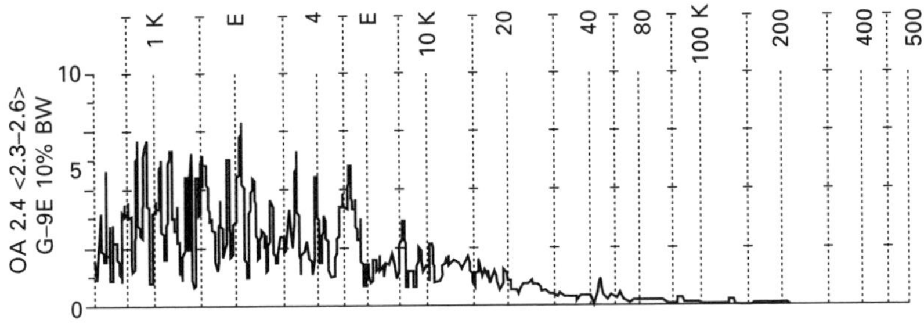

2.13 Frequency spectrum of 'A' bearing (outer race) in acceleration mode after 86 h of operation.

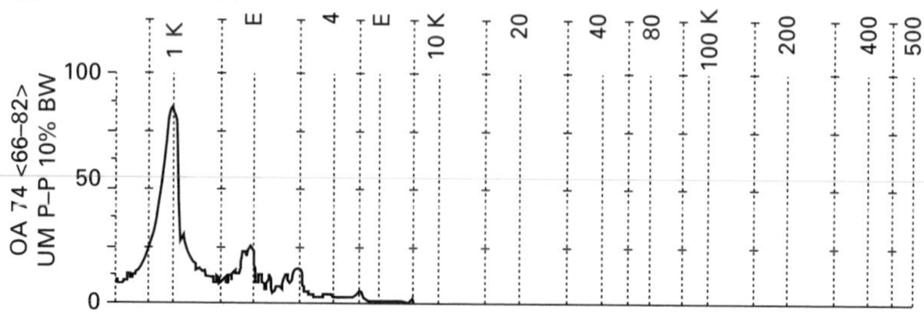

2.14 Frequency spectrum of 'A' bearing (outer race) in amplitude mode after 150 h of operation.

2.7.3 Analysis of kurtosis values of bearings under different stages of operation

Kurtosis value of bands K_1 to K_5 and K_A

The analysis of values of acceleration signals shown in Fig. 2.8 and 2.9 for 'A' and 'B' bearings indicates that the kurtosis values (K_1 to K_5) for the bearings vary between 3 and 11. Up to the third stage of operation the kurtosis value of 'A' bearing was higher than 'B'. But in the fourth and fifth stages, the values of K_1 to K_5 for 'B' bearing increased (Fig. 2.8 and 2.9). This indicates the initiation of deterioration in the 'B' bearing, and may be attributed to the presence of voids in the bearing material. The visual inspection of the bearing after the test indicates surface deterioration.

The K_A factor, used for the trend monitoring for the 'A' bearing, varied between 30 and 46. But at the fifth stage of operation, this stabilized to 30 as against the 'B' bearing, which shows the high rate of increase and approaches to 35 beyond the fifth stage of operation. This indicates the rate of change of the 'B' bearing condition during the course of operation.

Vibration velocity mm s^{-1} under different operating stages

The vibration velocity is calculated in the frequency ranges of 10 to 1000 kHz and is measured in mms^{-1} rms units. The comparison of the vibration velocities of the 'A' and 'B' bearings indicates that in both cases the vibration velocity has not changed during operation from first to fifth stages. The vibration velocity was slightly higher in the 'B' (5–6mms^{-1}) than the 'A' bearing (3–4mms^{-1}) as shown in Fig. 2.9. The system being the same, this may be attributed to the distortion of the 'B' bearing.

Acceleration levels

The values of G_1 to G_5 for 'A' bearing are more than that of 'B' bearing up to the third stage of operation, but in the fourth and fifth stage the values of

G_1 to G_4 for 'B' bearing exceed the 'A' bearing (Fig. 2.9 and 2.10). Also, the 'B' bearing shows the increasing trend from first to fifth stage for G_1 to G_5 bands compared with the 'A' bearing where the values of G_1 to G_5 are stabilized with run time. Initially, higher *g* values in a different frequency range indicate the irregularities of the rolling surfaces of new bearings, which get smoothened after a few hours of operation. The 'A' bearing shows better performance at stipulated higher operating parameters as compared with the 'B' bearing, which shows the deteriorated performance as shown in the Fig. 2.9 and 2.10. This is also in line with the analysis of K_A, K_1 to K_5 bands.

2.8 Overall vibration levels of bearings

The overall vibration levels of an 'A' bearing varied from 4 to 14 ms^{-2} during the run time under different operating stages but in the fifth stage the overall vibration level reduced and stabilized at 8 ms^{-2} (Fig. 2.11). In contrast, the overall vibration level of the 'B' bearing increased in the fifth stage and did not tend to stabilize. However, up to the fourth stage the vibration level varied between 3.5 and 5.5 ms^{-2}. This shows that the 'B' bearing condition deteriorated under stipulated higher operating parameters. Figures 2.12 to 2.14, showing the frequency spectrum of the 'A' bearing after 86 and 150 h of operation, indicate that there was no change in vibration amplitude with run time and hence the bearing condition.

2.9 Predicted life of bearings

Predicting the life of bearings is very difficult because of several factors affecting the bearing's performance, i.e. load, speed, lubrication, quality, etc. Using the kurtosis method of bearing damage detection, the rate of change of the K_A factor over a period of time is calculated, which gives the rate of damage that will best indicate the approximate life of the bearing. The following relation is used to predict bearing life *L*:[11]

$$L = \frac{50 - K_A(T_2)}{dK_A/dT} \qquad [2.4]$$

where dK_A is the change of K_A factor within the time interval T_1 and T_2, and 50 is taken as the value of K_A when the bearing is fully deteriorated.[11]

For the 'A' bearing K_A (T_2) was taken as 32 and dK_A/dT approximately as 1 as against K_A (T_2) 36 and dK_A/dT as 2 for the 'B' bearing (Fig. 2.9). This shows that the 'B' make bearing has approximately one-third the life of that of the 'A' bearing under higher operating parameters.

2.10 Comparison of performance of bearings

From the above data analysis, it is evident that the 'B' bearing shows the better performance up to the medium parameters of operation than the 'A' bearing. But at the higher stipulated parameters of operation (at an outer race temperature of 75°), i.e. in the fifth stage of operation, bearing behaviour deteriorates. In fact, from the fourth stage onwards, incipient damage in the 'B' bearing is initiated. This clearly indicates that at the stipulated higher operating parameters, which are a prerequisite for operation of these bearings, the 'B' bearings do not meet the acceptable functional requirements. As predicted, under these conditions the life of the 'B' bearing is approximately one-third that of the 'A' bearing. This has been confirmed by the field trials.

The difference in the performance of the bearing may be attributed to the following:

- Difference in number of rolling-elements in the bearings. This explains the higher load on the rollers in the loaded zone of the 'B' bearing (12 rollers) as compared with the 'A' bearing (13 rollers). During bearing operation, this probably gives better performance for 'B' bearings up to medium load conditions (Fig. 2.4, 2.7 and 2.11).
- There may be difference in the quality of material (with voids/impurities) of both bearings, which lead to initial scoring on the rolling surfaces of the 'B' bearing under higher operating parameters.
- Geometrical inaccuracy of the components, quality of manufacture and surface finish of the 'B' bearing may be inadequate, thus leading to the detection of initial incipient damage on its surface on inspection after test.
- Difference in quality control norms and use of precise instrumentation might lead to a difference in net quality of 'A' and 'B' bearings.

2.11 Conclusions

From the analysis of the test data discussed in this chapter, the following conclusions are drawn:[13]

- The stabilized temperature and rate of change of bearing temperature at higher operating parameters are better indications of bearing quality.
- The change in shock pulse data with run time indicates the bearing health.
- The rate of change of kurtosis, velocity and acceleration values in different bands indicate the process of deterioration of a test bearing during operation.
- The rate of change of kurtosis factor K_A with time as derived from test data can be used to predict the life of a test bearing under stipulated operating parameters.
- The life expectancy of 'B' bearings as derived from functional performance analysis is one-third that of 'A' bearings under stipulated operating conditions and matches closely with the field data.

- The lower life expectancy of 'B' bearings may be attributed to the quality and material of the bearings.

2.12 References

1 Markert, F.J. and Schweinfurt, S.K.F., 'Functional and Acceptance Testing in Rolling Bearing Development', *Ball Bearing J.*, **224**, 28–32, 1985.

2 Prashad, H., 'Performance Evaluation of Rolling Element Bearings', National *Conference of Mach. Mech.*, IISc, Bangalore (India), 1985.

3 Witte, D.C., 'Forecasting the Operating Temperature of Bearings', D K 621–822, 6536:721, Machine-31 (Din), 2/77, pp. 80–83.

4 Goldberg, D.D. *et al.*, 'Selection of Test Acceleration Coefficients for Bearing Units of Electrical Machines', *Electrotekhnika*, **55**, 10, 34–36, 1984.

5 Taylor, S.C., 'Noise in Bearings', *J. Eng. Mater. Des.*, February 1979.

6 Mathew, J. and Alfredson, R.J., 'The Condition Monitoring of Rolling Element Bearings using Vibration Analysis', ASME *J. Vibr., Acoust., Stress, Reliab. Des.*, **106**, 447–453, July 1984.

7 Prashad, H., 'Condition Monitoring of Anti-friction Bearing', *J. Inst. Eng.* (*I*), **69**, 65–74, 1989.

8 Prashad, H., *et al.*, 'Diagnostic Monitoring of Rolling Element Bearings by High Frequency Resonance Technique', *ASLE Trans.*, **28**(4) 439–448, 1985.

9 Prashad, H., 'The Effect of Cage and Roller Slip on the Measured Defect Frequency Response of Rolling Element Bearings', *ASLE Trans.*, **30**(3), 360–367, 1986.

10 Betthel, K., 'The Shock Pulse Method for Monitoring the Condition of Anti Friction Bearings', SPM Catalogue, 1977.

11 *Kurtosis Technical Handbook*, Condition Monitoring Ltd, 1982.

12 *Ball Roller Bearing Eng., Indus. Eng. J.*, FAG, EA 10–12, 1983.

13 Prashad, H., 'Functional Performance of Roller Bearings for Acceptance in Routine Applications', *J. Inst. Eng. India, Part MC, Div., Mech. Eng.* **70**, 105–113, 1995.

3 Cage and roller slip of rolling-element bearings

3.1 A general review

This chapter explains the effect of cage and roller slip on the measured defect frequency response of rolling-element bearings using the high-frequency resonance technique (HFRT) as a surveillance module. The difference in theoretically evaluated and experimentally determined defect frequencies has been used to identify percentage cage and roller slip in various bearings under no-load and load conditions at different speeds of operation. Various defect frequencies and their energy levels have been used to monitor the severity of defects and complex response characteristics/interactions of rolling-element with inner and outer rings of the bearings.

It is shown that negative cage and roller slip is predominant at moderate speed under no-load and load operation. Positive slip is significant at high-speed under no-load operation. Negative cage and roller slip is indicated by the presence of outer-race linear defect frequency along with rolling-element defect frequencies. Energy levels of identified defect frequencies are greater under no-load than under load for new bearings at all speeds of operation. However, for defective bearings, this pattern is reversed at rated and high speeds of operation. In general, percentage negative cage slip is found to be more than the corresponding roller slip. Under load at rated speed, cage and roller slip is minimal for new bearings. However, slip may still be identified by defect frequency response.

3.2 Introduction

In most roller bearing applications, operating conditions are such that the cage and roller motions are essentially epicyclical. However, in some situations of high speeds and light loads, several investigators have reported roller bearing cage slip, in which cage and roller assembly travel at speeds lower than predicted from epicyclical considerations.[1] Smith reported considerable cage slip in the main shaft bearings operating at high speeds under light loads.[2] Dowson and Higginson[3] suggested that considerable cage and roller slip could be expected in lightly loaded bearings operating at moderate speeds; whereas under heavier loads when the conditions would be elastohydrodynamic, the roller-bearing motion would be essentially epicyclical. However, Harris,[4] by applying analytical methods based on elastohydrodynamic considerations, has shown that significant departure from epicyclical motion can exist in heavily loaded bearings operating under elastohydrodynamic considerations, which is inconsistent with the conclusions of Dowson and Higginson for that region.

The present study explains the cage and roller slip phenomenon and its effect on defect frequency response under load and no-load conditions at various operating speeds. Diagnostic monitoring using HFRT has been used to study the above phenomenon in rolling-element bearings.

3.3 Characteristic defect frequencies

The contact between the mating surfaces in the bearing is periodic and impulses occur at regular intervals. The frequency of occurrence of such impulses is taken as the characteristic frequency corresponding to the defects in the bearing elements. The spectral identification of these frequencies is taken as the basis for the diagnostic monitoring. The characteristic defect frequencies are derived theoretically from kinematics analysis of the defect impacting action of he bearings. The defect frequencies are given in references 5–7 and worked out in Chapter 1.

These frequencies are approximate since they are affected by slipping of bearing elements. Under normal load conditions, slip is insignificant but, at high speeds and light loads or at no-loads, slip may be quite high.[8,9] Large positive cage slip (so-called skidding) may be detrimental to the service life of bearing. It can frequently cause surface distress on the inner race, like smearing.[10] The slip phenomenon thus occurring in the bearings has been correlated with the defect frequencies response.

3.4 Experimental procedure

Various bearings of type NU 330 were tested under load and no-load conditions up to the rated speed of 2250 rpm and high speed of 2750 rpm. The drive motor assembled with bearings under test is coupled to a similar motor through a rigid coupling. The test set-up is shown in Fig. 1.1 of Chapter 1. The coupled motor acts as a generator and the drive motor acts as a motor under test. Speed variation is achieved by varying voltage and current input to the motor and speed is monitored by a magnetic pick-up and displayed on a digital speed indicator. No-load and load operations of bearings are considered as tested under uncoupled and coupled conditions, respectively.

In all tests, a vibration accelerometer with a magnetic base was mounted radially on the bearing outer race. The signals from pick-up were recorded on a tape recorder in acceleration mode to carry out signature analysis. The temperatures of the bearing outer race and housing were also monitored by spring-loaded thermocouples.

3.5 Spectral analysis technique

Frequency analysis of the recorded signatures is analysed using a real-time analyser (RTA) to determine the carrier (resonant) frequencies most strongly modulated by the defects frequencies. The recorded accelerometer signals are then demodulated about these frequencies by the envelope-detector and re-recorded on another magnetic tape recorder for evaluation of bearing defect frequencies and their energy levels using RTA.

The envelope-detector works on the principle of mixing two signals and detecting the wanted signal by super-heterodyning. The system consists of voltage-controlled oscillator (VCO), mixer, counter, band-pass filter, detector and variable cut-off low-pass filter.

The carrier frequency, as detected by the RTA, is set with the VCO and the counter. The linearity of the VCO is better than 0.01% and stability is about 10 Hz in 60 kHz with a resolution of 1 Hz.

The set carrier frequency signal from the VCO and the recorded bearing signals are fed to the mixer and output is passed through a band-pass filter with a 2.2 kHz bandwidth. Signals thus obtained are then filtered with variable cut-off low-pass filter before recording on the tape recorder.
Brief technical particulars of the detector are as follows:

- Frequency range : 3 to 60 kHz
- Frequency setting accuracy : ± 1 per cent
- Cutoff frequencies : 100, 200, 500, 800 and 1000 Hz
- Accuracy : ± 3 per cent
- Fall-off rate : 24 dB/octave
- Dynamic range : 40 dB

The accelerometer, having a charge sensitivity of 10 pC g^{-7}, is used within the 0.2 to 12 000 Hz frequency range. The low-frequency cut-off of the accelerometer is determined by preamplifier and environmental conditions. The charge preamplifier is used with the accelerometer as a charge source and produces an output voltage proportional to the change in input voltage. A large amount of capacitive feedback is used to obtain a very high preamplifier input capacitance and, therefore a very long accelerometer connection cable can be used without the shunt capacitance altering the lower limiting frequency for measurement.

3.6 Determination of defect frequencies and energy levels

3.6.1 Theoretical

Various defect frequencies and cage and rolling-element frequencies (f_c, f_b) at different speeds of operation are theoretically determined for the NU 330 bearings as per the formulae given in Section 1.3 of Chapter 1 (refer to Section 3.11). The kinematics of the bearing are given as $D = 238$ mm, $d = 45$ mm, $\alpha = 0$ and $N = 14$.

Average experimental values of cage and rolling-element frequencies (f_{c1av}, f_{b1av}) are determined from the various experimentally evaluated defect frequencies as per the following formulae which are derived using equations [1.1], [1.2] and [1.10] (see Section 3.11):

$$f_{c11} = \frac{f_{RW1(I)}}{3 + 2D/d} \quad [3.1]$$

$$f_{c12} = \frac{f_{RW1(II)}}{1 + 2D/d} \quad [3.2]$$

By using equation [1.4], the following is evident (see Section 3.11):

$$f_{c13} = \frac{(f_{ORDFL_1})}{N} \quad [3.3]$$

Using equations [1.9], [1.1] and [1.2] gives (see Section 3.11):

$$f_{c14} = \frac{df_{RRS_1}}{2(D + d)} \quad [3.4]$$

and

$$f_{clav} = \frac{f_{c11} + f_{c12} + f_{c13} + \ldots + f_{clm}}{m} \quad [3.5]$$

By using equations [1.1] and [1.2], it is evident that (see Section 3.11):

$$f_{blav} = \frac{f_{clav}\,(D + d)}{d} \quad [3.6]$$

Percentage cage slip is calculated as:

$$\Delta f_c = \frac{f_c - f_{clav}}{f_c} \times 100 \quad [3.7]$$

Similarly, percentage roller slip is calculated as:

$$\Delta f_b = \frac{f_b - f_{blav}}{f_b} \times 100 \quad [3.8]$$

3.6.2 Experimental

The test data on three bearings of NU 330 type are reported: 'A' (new), 'A' (defective) and 'B' (defective). Bearings 'A' and 'B' are of different makes. The bearing 'A' (defective and 'B' (defective) have known defects; bearing 'A' (defective) has a through crack in the inner race in the axial direction and 'B' (defective) has a score on the rollers. 'A' and 'B' bearings were tested after filling a fixed quantity of EP (extreme pressure) grease (viscosity 127.33 cSt at 40 °C and 10.43 cSt at 100 °C).

Based on the calculated resonant frequencies, as the formulae given in Section 1.4, and the predominant frequencies existing in the raw signature, the significant centre frequencies were established. Then detailed signature analysis of bearings were carried out by using envelope-detector/HFRT, with the significant centre frequencies ranging from 0 to 10 kHz at various speeds of operation under coupled mode. The frequency analysis was made and the plots were drawn in the range of 0 to 500 Hz. The number of averages in the analysis was taken as 64. Energy levels were calculated for significant defect frequencies, based on the signal conditioner sensitivity (mV $s^2 m^{-1}$) rms and input demodulation level (dB) for the respective frequencies. Energy levels are arbitrarily used as the acceleration levels in m s^{-2} rms times the unit total amplitude (voltage) ratio of the demodulated signals comprising the defect frequencies. This does not carry any reference to other standard system of units.

Theoretical and experimental values of cage and roller frequencies along with percentage cage slip Δf_c and roller slip Δf_b at various operating parameters are given in Table 3.1. Table 3.2 indicates various significant defect frequencies and their energy levels for 'A' (new) and 'A' (defective) bearings under 'no-load' and 'load' operation at 1000, 1800, 2250 and 2750 rpm along with percentage cage and roller slip. The analysis of 'B' (defective) bearing at 1800, 2250 and 2750 rpm is given in Table 3.3. A typical plot showing significant defect frequencies under 'load' operation is shown in Fig. 3.1 at 2250 rpm of operation for the 'A' (new) bearing. Similarly, Fig. 3.2 shows a demodulated spectrum of the 'A' (defective) bearing under 'no-load' at 2750 rpm. These plots are representative of the frequency spectrum. Raw signatures without demodulation do not indicate defect frequencies.

3.7 Results and discussion

3.7.1 Cage and roller slip

From Table 3.1, it is apparent that percentage cage slip (Δf_c) and percentage roller slip (Δf_b) vary from –7.6 to 5.6 and –5.8 to 7.1, respectively, in the complete range of operation for the testing bearings. Negative Δf_c and Δf_b at moderate speed under 'no-load' and 'load' operation change to positive Δf_c

Table 3.1 Percentage cage slip (Δf_c) and roller slip (Δf_b) of bearings under load and no-load conditions at different rpm of operation

Speed	Theoretical frequency	Experimental data											
		A (new)				A (defective)				B (defective)			
rpm	f_c (Hz)	No-load		Load		No-load		Load		No-load		Load	
		f_{clav}	Δf_c	f_{clav}	Δf_c	f_{clav}	Δf_c	f_{clav}	Δf_c	f_{clav}	Δf_c	f_{clav}	Δf_c
(1)	(2)	(3)	(4)	(5)	(6)	(7)	(8)	(9)	(10)	(11)	(12)	(13)	(14)
1000	6.67	7.18	–7.6	7.18	–7.6	6.93	–3.9	6.62	0.75				
1800	12.0	12.27	–2.27	12.12	–0.90	12.29	–2.4	12.14	–1.19	12.21	–1.7	12.17	–1.4
2250	15.2	15.40	–1.3	15.29	–0.56	15.02	1.17	14.87	2.17	15.25	–0.3		
2750	18.3	17.27	5.6			17.85	2.5			19.26	–5.2		
rpm	f_b (Hz)	f_{blav}	Δf_b	f_{blav}	Δf_b	f_{blav}	Δf_b	f_{blav}	Δf_b	f_{blav}	Δf_b	f_{blav}	Δf_b
1000	42.5	45.15	–5.8	45.15	–5.8	43.59	–2.5	41.6	2.1				
1800	76.5	77.18	–0.8	76.23	~0.3	77.30	–1	76.36	~0.2	76.80	–0.4	76.55	–0.06
2250	95.8	96.87	–1.1	96.17	–0.4	94.47	1.4	93.53	2.3	95.92	–0.13		
2750	116.8	108.6	7.1			112.28	4.0			121.15	–3.6		

Table 3.2 Frequency analysis of 'A' (new) and 'A' (defective) bearings under no-load conditions at differing rpm of operation

Bearing operation	A (new)				A (defective)				Theor. defect machining		ΔEL	ΔEL
	No-load		Load		No-load		Load					
Δf_c, Δf_b	−7.6 Fr (Hz)	−5.8 EL	−7.6 Fr (Hz)	−5.8 EL	−3.9 Fr (Hz)	−2.5 EL	0.75 Fr (Hz)	2.1 EL	Fr (Hz)	Defects	A (new)	A (defective)
					84	121.6			85	RRS_1		
1000 rpm	96	146	96	44			90	88.5	91.7	RW_1(I)	102	
	102	166.5	102	28.3	100	121.6			93.4	$ORDFL_1$	138.2	
							124	74	122	$IRWL_1$(II)		
							140	75	139	$IRDFL_1$		
Δf_c, Δf_b	−2.27	−0.8	−0.90	~0.3	−2.4	−1	−1.19	~0.2				
	142	170.5	140	40.9					141	RW_1(II)	129.6	
					152	383			153	RRS_1		
1800 rpm	172	152	170	80.9	174	376	170@	128.4	168	$ORDFL_1$	71.1	247.6
							170@		165	RW_1(I)		
			210	45.9			220	63	220	$IRWL_1$(II)		
			240	35.4	254	423	250	53.6	250	$IRDFL_1$		369.4
			340	42.8					336	$ORDFL_2$		
Δf_c, Δf_b	−1.3	−1.1	−0.56	−0.4	1.17	1.4	2.17	2.3				
	178	646							180	RW_1(II)		
					190	767	190	788	191.6	RRS_1		−21
			214@		204	749	202	1176	210	RW_1(I)		−427
2250 rpm	216	733	214@	296.5					214	$ORDFL_1$	436.5	
	280	159			280	691	283	1246	280	$IRWL_1$(II)		−555
	318	189			318	1333	314		318	$IRDFL_1$		

Table 3.2 Continued

Bearing operation	A (new)				A (defective)				Theoring defect machining		ΔEL	ΔEL
	No-load		Load		No-load		Load					
Δf_c, Δf_b	5.6	7.1			2.5	4						
	Fr (Hz)	EL	Fr (Hz)	EL	Fr (Hz)	EL	Fr (Hz)	EL	Fr (Hz)	Defects	A (new)	A (defective)
2750 rpm	200	199			202	1078			215	RW_1(II)		
					248	968			251.1	RW_1(I)		
	326	95			340	713			336	$IRWL_1$(II)		
					388	1650			382	$IRDFL_1$		

Fr = frequency, EL = energy level ΔEL = difference in energy level under no-load and load operation.

Table 3.3 Frequency analysis of 'B' (defective) bearing under no-load and load conditions at differing rpm of operation

Operation	No-load		Load		Theoretical defect matching		ΔEL
Δf_c, Δf_b	–1.7	–0.4	–1.4	0.06			
1800 rpm	Fr (Hz)	EL	Fr (Hz)	EL	Fr (Hz)	Defects	–
	141.25	140			141	RW_1(II)	
			151.25	195	153	RRS_1	–55.0
	171.25	105	171.25	168.75	168	$ORDFL_1$	–63.75
Δf_c, Δf_b	–0.3	–0.13					
2250 rpm	175.0	104			180	RW_1(II)	
	212.5	192			214	$ORDFL_1$	
	387.0	69			383.2	RRS_2	
	426.0	91			428.0	$ORDFL_2$	
Δf_c, Δf_b	–5.2	–3.6					
2750 rpm	230	862			215	RW_1(II)	
	261.25	502			256	$ORDFL_1$	

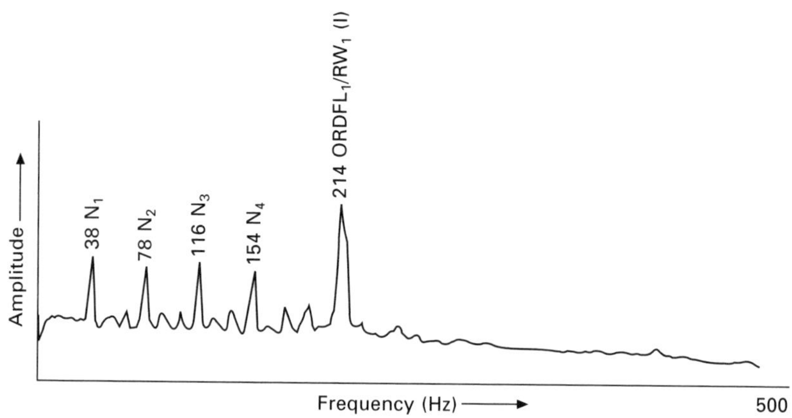

3.1 Demodulated frequency spectrum of A (new) bearing at 2250 rpm under load operation.

and Δf_b at high speed and 'no-load'. In general, negative Δf_c is more than the corresponding Δf_b, and positive Δf_c is less than the corresponding Δf_b. This may be due to the cage friction forces at different operating parameters.

At rated speed (2250 rpm) under load, cage slip and roller slip in new bearings become negligible (Δf_c = –0.56, Δf_b = –0.4) and bearings show more or less a no-slip condition. This may be due to elastohydrodynamic operation of the bearing.

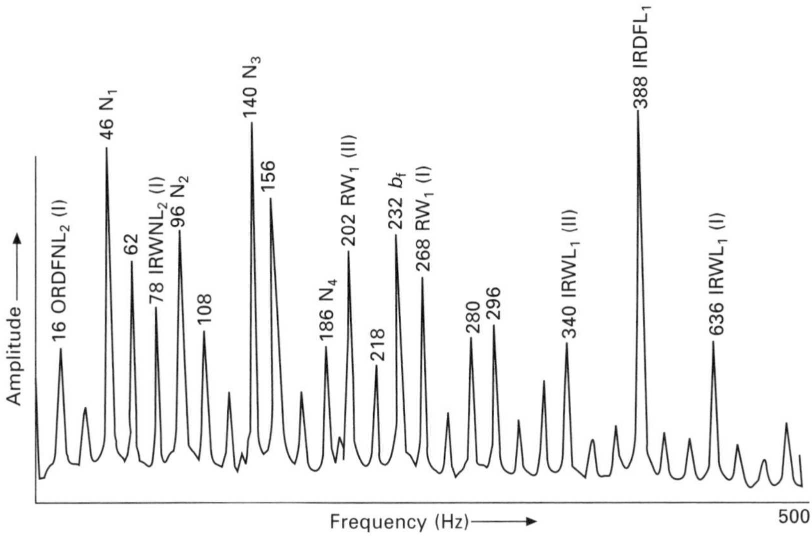

3.2 Demodulated frequency spectrum of A (defective) bearing at 2750 rpm under no-load operation.

Positive cage and roller slip shows the phenomenon of sliding of rollers rather than rotation. A rise in temperature, reduction in lubricant viscosity and bearing clearance may reduce the values of Δf_c and Δf_b more in the 'A' (defective) bearing ($\Delta f_c = 2.5, \Delta f_b = 4.0$) than in the 'A' (new) bearing ($\Delta f_c = 5.6, \Delta f_b = 7.1$) as shown in Table 3.1.

3.7.2 'A' (new) bearing

It is evident from Table 3.2 that the 'A' (new) bearing at different rpm under 'no-load' and 'load' operation exhibits certain defect frequencies with varying energy levels. Energy levels of defect frequencies under 'no-load' are much higher than under 'load' conditions and also increase with speed of operation. The difference in energy levels for f_{ORDFL1} defect is 138.2, 71.1 and 436.5 at 1000, 1800 and 2250 rpm of operation, respectively, under 'no-load' and 'load' conditions. Under these conditions, the difference in energy levels for $f_{RW1(I)}$ defect is 102 at 1000 rpm and for $f_{RW1(II)}$ defect is 129.6.

The phenomenon of negative cage slip – when motion is faster than its epicyclical value – is observed by the presence of f_{ORDFL} and f_{RW} under 'no-load' and 'load' conditions at moderate speeds. At rated speed (2250 rpm) under load $f_{ORDFL1}/f_{RW1(1)}$ is also observed. However, Δf_c (–0.56) and Δf_b (–0.40) are negligible. The theoretical values of f_{ORDFL1} and $f_{RW1(1)}$ are very closely matching with maximum difference of 4 Hz at 2750 rpm. These are found to be overlapping (represented by the @ symbol) on each other in the defect frequencies response of different bearings at certain operating parameters (in 'A' (defective) bearing under 'load' at 1800 rpm and 'A' (new) under 'load' at 2250 rpm).

The apparent drop in percentage cage slip Δf_c from –7.6 at 1000 rpm to –0.90 at 1800 rpm under 'load' and from –2.27 at 1800 rpm to –1.3 at 2250 rpm under 'no-load' with corresponding change in Δf_b (–5.8 to 1.1) suggests the contact of roller edges with inner race. This initiates various rolling-element and inner-race defects frequencies at 1800 rpm (load) and 2250 rpm (no-load) as shown in Fig. 3.1.

Increase in speed under 'no-load' generally increases the oil film thickness between inner ring and rollers, causing a tangential driving force, which is inversely proportional to the film thickness, and tends to decrease it. However, the radial outward movement of the rollers due to the increase in film thickness results in a larger frictional drag force between the rollers and outer race. This phenomenon causes surface distress on the outer race and gives rise to f_{ORDFL} and f_{RW} with higher energy levels under 'no-load' at different speeds as against operation under 'load'.

To re-establish equilibrium between drag and driving tangential forces, the cage travels more slowly than its epicyclical value as shown by $\Delta f_c = 5.6$ and $\Delta f_b = 7.1$ at 2750 rpm under 'no-load' (Table 3.1). The change of negative slip at 2250 rpm to positive slip – when cage motion is slower than its epicyclical value – at 2750 rpm may be related to the decrease in lubricant viscosity, speed parameter and film thickness. This causes surface distress on the inner-race defect frequency accompanied by a roller defect frequency at 2750 rpm under no-load operation. On the contrary, at 1000 rpm under 'no-load', energy levels of 166.5 and 146.0 as against 28.3 and 44.0 under 'load' for f_{ORDFL1} and $f_{RW1(1)}$, respectively, suggest the negative slip and higher surface distress on the outer race and rolling-elements.

3.7.3 'A' (defective) bearing

The difference in energy levels of various defect frequencies is much higher under 'no-load' than under 'load' at 1000 and 1800 rpm similar to the 'A' (new) bearing. In contrast, at 2250 rpm of operation the 'A' (defective) shows higher energy levels under 'load' as against those under 'no-load' operation (Table 3.2). This illustrates that the deformation and excitation in the defective bearing is more predominant under 'load' than under 'no-load' conditions at higher speed, because of the existence of a through axial crack in the inner race. The presence of various inner-race defect frequencies ($f_{IRWL1(II)}$, f_{IRDFL1}) at 1800 and 2250 rpm with significant energy levels under 'no-load' and 'load' conditions illustrates the existence of known defects in the 'A' (defective) bearing.

The energy levels of defect frequencies increase with speed under 'load' and 'no-load' operation. Under load, energy levels for defects related to $f_{RW1(I)}$ and $f_{IRWL1(II)}$ increase from 88.5 and 74 at 1000 rpm to 1176 and 1246 at 2250 rpm, respectively. Similarly under 'no-load', energy levels at various defects such as f_{IRDFL1}, f_{RRS1} increase from 423 and 383 to 1333 and 767, respectively, as the speed changes from 1800 to 2250 rpm (Table 3.2).

With speed, the centrifugal force on the bearing increases. This tends to increase the inner-ring diameter because of a through axial crack in it. This may decrease the bearing clearance and increase contact of rolling elements with inner race as compared with the outer race, resulting in surface distress on the inner race and an exorbitant increase in energy levels of the inner race (1650 for f_{IRDFL}) and rolling-elements (1078 for $f_{RW1(II)}$) for defect frequencies at 2750 rpm. The decrease in radial clearance also results in decreasing the cage and roller slip (Δf_c = 2.5 and Δf_b = 4.0) more significantly for 'A' (defective) bearing as compared with 'A' (new) bearing (Δf_c = 5.6 and Δf_b = 7.1) at 2750 rpm under 'no-load'. This is similar to the observations made by Kaido and Doi.[10]

The presence of positive cage and roller slip (Δf_c = 1.17 and Δf_b = 1.41) at 2250 rpm and 2750 rpm (Δf_c = 2.5 and Δf_b = 4) under 'no-load' indicates the presence of inner-race defect frequencies accompanied by roller defect frequencies similar to the new bearing at 2750 rpm of operation. The same is also shown at 1000 and 2250 rpm under 'load'. However, at moderate speeds (1000 to 1800 rpm) at 'no-load', the defective bearing shows negative slip similar to the new bearing.

3.7.4 'B' (defective) bearing

'B' (defective) bearings, with score marks on the rollers, show defect frequencies for roller waviness and roller rough spots under 'no-load' and 'load' operation as shown in Table 3.3. Owing to the existing defects in the bearing, energy levels of roller and outer-race defects are higher by 55.0 and 63.75 under load as compared with no-load operation at 1800 rpm. The bearing also shows an increase in energy levels of defect frequencies with speed. A defective bearing indicates negative cage and roller slip at all speeds of operation and indicates f_{ORDFL} and f_{RW}. Negative cage and roller slip is detected because of the low temperature rise and higher film thickness in the bearing, as the bearing was operated only under 'no-load' at 2250 and 2750 rpm for a short duration.

3.8 Comparison of bearings

Negative cage and roller slip may be identified by the presence of f_{ORDFL} and f_{RW}. It is more predominant at moderate speeds under 'no-load' and 'load' conditions. At higher speed and 'no-load', positive cage and roller slip become more significant and it is identified by the presence of various inner race and roller defect frequencies. Rise in bearing temperature and other built-in defects affect the negative and positive slip phenomenon in the bearings.

The 'A' (defective) bearing with severe inner-race defects shows higher energy levels at N_1 to N_3 (varying between 1520 to 1650) at 2750 rpm apart from the energy levels (713, 1650) of the inner race defects. The burst of energy levels of the various inner race defect frequencies and shaft rotational frequencies reflect the movement of the defect into and out of the loaded zone. The burst of energy levels is much lower for the 'A' (new) bearing under all conditions of operation as compared with the 'A' (defective) bearing. Similarly, the 'B' (defective) bearing having score on the rollers shows much higher energy levels for rotating speeds and its components as compared with the 'A' (new) bearing.

It has been noted that with an increase of speed and load, linear defects become more predominant in the spectrum. The source of non-linear defects in the bearing are significant, particularly at low speed under 'no-load' operation, and may be confined to vibrations of rolling-elements, eccentricity and low-order waviness of races. Plots at higher speed under 'no-load' are more complex than under 'load' operation (Fig. 3.1 and 3.2). These are due to the presence of different harmonics of defect frequencies, rotating speed frequencies, and other complex response interactions at high speed under 'no-load' and low speed under 'load'.

3.9 Conclusions

The following conclusions can be drawn from the above study:[11]

- Bearings show negative cage and roller slip at moderate speed under 'no-load' and 'load' operation. This has been assessed by detection of roller and linear outer-race defect frequencies.
- Bearings indicate positive cage and roller slip at high speed under 'no-load' operation. This is identified by detection of inner race accompanied by roller defect frequencies.
- In general, percentage negative cage slip is more than the corresponding roller slip and percentage positive cage slip is less than the corresponding roller slip.
- Minimum cage and roller slip prevail at rated speed under load conditions. This may be due to complete elastohydrodynamic operation. The slip is still identified by defect frequency response.
- At high speed, defective bearings show exorbitant energy levels for defect frequencies, rotating speed frequency and its harmonics.
- In new bearings, energy levels of various defect frequencies are two to four times more under 'no-load' than under 'load' at different operating speeds. In defective bearings, this pattern is reversed at high speeds.
- Energy levels of various defect frequencies increase with speed.

Although not discussed explicitly, it was observed that temperature rise and thereby reduction in radial clearance affect the cage and roller slip.

3.10 References

1 Boness, R.J., 'Cage and Roller-Slip in High Speed Roller bearings', *J. Mech. Sci.*, **2**(2), 1969.
2 Smith, C.F., 'Some Aspects of the Performance of High Speed Lightly Loaded Cylindrical Roller Bearings', *Proc. Inst. Mech. Eng.*, **176**(227), 566, 1962.
3 Dowson, D. and Higginson, G.R., 'Theory of Roller Bearing Lubrication and Deformation', *Proc. Lubr. Wear Covn. (Inst. Mech. Eng.)*, London, 216, 1963.
4 Harris, T.A., 'An Analytical Method to Predict Skidding in High-speed Roller Bearings', *ASLE Trans.*, **9**(3), 229–241, 1966.
5 Prashad, H., Ghosh, M. and Biswas, S., 'Diagnostic Monitoring of Rolling-element Bearings by High Frequency Resonance Technique', *ASLE Trans.*, **28**(4) 439–448, 1985.
6 Sramek, B., 'Frequency Calculations for Ball Bearings', Preliminary note, Nicolet Scientific Corp., April, 1978.
7 Nishio, K. *et al.*, 'An Investigation of Early Detection of Defects in Ball Bearings by Vibration Monitoring', ASME Paper 79-DET-45, 1979.
8 McFadden, P.D. and Smith, J.D., 'Vibration Monitoring of Rolling-element Bearings by High-Frequency Resonance Technique – A Review', *Tribol. Int.*, **17**(1), 84, 1986.
9 Steward, R.M., 'Application of Signal Processing Techniques to Machinery Health Monitoring, Application of Time Series Analysis', paper available from Inst. of Sound and Vibration Research, University of Southampton, 16.1–16.23, April 1980.
10 Kaido, H. and Doi, Y., 'Cage Slip and its Effect on Damages in High Speed Roller Bearing', *Proc. JSLE Intl. Tribol. Conf.*, Tokyo, Japan, pp. 591–596, 8–10, July 1985.
11 Prashad, H., 'The Effect of Cage and Roller Slip on the Measured Defect Frequency Response of Rolling-element Bearings', *ASLE Trans.*, **30**(3), 360–366, 1987.

3.11 Appendix

$$f_c = \frac{0.5 f_s}{1 - (d/D)\cos\alpha} \quad \text{[A3.1]}$$

$$f_b = 0.5\, f_s (D/d)\, [1 - (d/D)^2 \cos^2\alpha] \quad \text{[A3.2]}$$

$$f_{bf} = 2\, n f_b = n f_s\, (D/d)\, [1 - (d/D)^2 \cos^2\alpha] \quad \text{[A3.3]}$$

$$f_{ORDFL} = nN f_c = \frac{0.5\, nN f_s}{1 - (d/D)\cos\alpha} \quad \text{[A3.4]}$$

$$f_{IRDFL} = nN(f_s - f_c) = \frac{0.5\, nN f_s}{1 + (d/D)\cos\alpha} \quad \text{[A3.5]}$$

$$f_{IRWL} = f_{IRDFL} \pm f_s \quad \text{[A3.6]}$$

$$f_{\mathrm{IRWNL}} = \frac{f_{\mathrm{IRDFL}}}{N \pm f_c} \quad \text{[A3.7]}$$

$$f_{\mathrm{ORDFNL}} = (n \pm 1) f_c \quad \text{[A3.8]}$$

$$f_{\mathrm{RRS}} = 2\, n f_b \quad \text{[A3.9]}$$

$$f_{\mathrm{RW}} = 2\, n f_b \pm f_c \quad \text{[A3.10]}$$

3.12 Nomenclature

α	contact angle
D	pitch diameter
d	diameter of rolling-element
f_s	shaft rotational frequency
f_c	cage rotational frequency (theoretical)
f_{c1m}	cage rotational frequency (experimental)
f_{c1av}	average cage rotational frequency (experimental)
f_b	rolling-element frequency (theoretical)
f_{b1m}	rolling-element frequency (experimental)
f_{b1av}	average rolling-element frequency (experimental)
Δf_b	percentage slip in rolling-element frequency
f_{bf}	rolling-element defect frequency
Δf_c	percentage slip in cage rotation frequency
f_{rr}	resonance frequency of race
f_{IRDFL}	inner race defect frequency (linear)
f_{IRDFL} (L_1, L_2)	inner race defect frequency (linear) for $n = 1$ and $n = 2$
f_{IRWL}	inner race waviness frequency (linear)
f_{IRWL} (L_1, L_2)	inner race waviness frequency (linear) for $n = 1$ and $n = 2$
f_{IRWNL}	inner race waviness frequency (non-linear)
f_{IRWNL} (L_1, L_2)	inner race waviness frequency (non-linear) for $n = 1$ and $n = 2$
f_{ORDFL}	outer race defect frequency (linear)
f_{ORDFL} (L_1, L_2)	outer race defect frequency (linear) for $n = 1$ and $n = 2$
f_{ORDFNL}	outer race defect frequency (non-linear)
f_{ORDFNL} (L_1, L_2)	outer race defect frequency (non-linear) for $n = 1$ and $n = 2$
f_{RW}	roller waviness frequency
f_{RRS}	roller rough spot frequency
f_{RRS} (S_1, S_2)	roller rough spot frequency for $n = 1$ and $n = 2$
f_{RW} (W_1, W_2)	roller waviness frequency for $n = 1$ and $n = 2$
m	number of frequencies (m = 1, 2, 3, ...)
n	1, 2, 3, ..., 9 (harmonics)

N	number of rolling-elements
r	radius of rolling-element
@	overlapping of experimental evaluated frequencies

Above frequencies with $L_1(I)$, $S_1(I)$, $W_1(I)$ correspond to $n = 1$ for first order and $L_1(II)$, $S_1(II)$, $W_1(II)$ to $n = 1$ for second order. Similarly, frequencies with $L_2(I)$, $S_2(I)$, $W_2(I)$ correspond to $n = 2$ for first order and $L_2(II)$, $S_2(II)$,$W_2(II)$ to $n = 2$ for second order.

4 Diagnosis and cause analysis of rolling-element bearings failure in electric power equipment

4.1 A general review

This chapter highlights the investigations pertaining to the diagnosis of rolling-element bearings of motors of electric power equipment that has failed due to the causes generally unforeseen during design and operation. However, in general, the diagnosis of the failure of the bearings has been well established in the literature. The causes unforeseen by failure diagnosis have been ascertained in this study. These generally unforeseen causes are found to allow the passage of an electric current through the bearings of the motors, causing them to deteriorate.

The vibration and shaft voltage data and the characteristics of the lubricant used in various rolling-element bearings were analysed. The cause of bearing failure in electric power equipment was diagnosed.

4.2 Introduction

4.2.1 Causes of shaft voltages and flow of current through bearings

There is a phenomenon called shaft voltages that exists in electrical machines. It causes a flow of current through the bearings, depending on the resistance of the bearing circuits. This has been discussed in detail.[1-15]

The flow of current through the bearing causes a magnetic flux to develop, which closes in the circumference over the yoke and induces the voltage on the shaft as the machine rotates. This results in a localized current at each bearing rather than a potential difference between the shaft ends. A current path, however, along the shaft, bearings and frame results in a potential difference between the shaft ends.[3]

At a certain threshold voltage, depending on the resistivity of the lubricant and operating conditions, a current flows through the bearing.[4] Thus, the flow of circular current in the inner race leaks through the rolling-elements to the outer race by following the path of least resistance and establishes a field strength leading to the development of magnetic flux on the track surface of races and rolling-elements.[5]

Studies have been carried out by various authors on the causes and control of electrical currents in bearings,[6] the flow of current through lubricated contacts,[7] the effect of electrical current on bearing life,[8] the effect of operating parameters on an impedance response,[4] and the deterioration of lubricants used in non-insulated bearings.[10–12] Surveys of the failure of rolling-element bearings indicate various causes, including that of failure due to corrugations.[13,14,16] The mechanism of formation of the corrugation pattern on the track surfaces and related investigations on roller bearing surfaces have also been carried out.

This case study deals with investigations pertaining to generally unforeseen causes that lead to the premature failure of rolling-element bearings in the motors of electric power equipment.

4.3 Bearing arrangement and nature of bearing failure

The motors have a rated capacity of 2100 kW and operate at 1494 rpm. They have a triple bearing arrangement. The motors are used to drive primary air fans.

On the non-drive end (NDE) of the motor, bearing type NU 228 is used and is insulated. On the drive end (DE) of the motor, roller bearing type NU 232 and ball bearing type NU 6326 are used for taking both radial and axial loads simultaneously. All the bearings are grease lubricated and the motors are designed for continuous operation with periodic relubrication. Bearings also have instruments for continuous measurement of the temperature during operation. Figure 5.1 in Chapter 5 shows the schematic diagram of the NDE bearing arrangement in the motor.

After commissioning of a few motors, bearings were found to be failing prematurely. The nature of the failure was the formation of corrugations and flutings on the roller tracks of the races as well as corrosion on the raceways and rolling-elements, irrespective of the make of the bearings used. However, the depth of corrugations and degree of development of corrugations varied. Figure 5.2 in Chapter 5 shows a photograph of corrugations and corrosion on the inner race track surface of ball bearing 6326 due to passage of current. The grease was also found to be blackened.

4.4 Investigations, observation of failures and data collection

The investigations pertaining to shaft voltage, vibration and shock pulse levels were carried out on various motors both on NDE and DE bearings.

4.4.1 Measurement of shaft voltages

The shaft voltages were measured between non-drive ends and the drive end shaft, drive end shaft to the ground, and non-drive end shaft to the ground. Table 4.1 shows shaft voltage data after removing the grounding brush. Table 4.2 indicates frequency analysis of different shaft voltages after spectrum analysis of recorded shaft voltages on tape recorder using shaft probes. Shaft voltages were also measured on retaining the grounding brush and keeping the brush intact (Fig. 5.1).

4.4.2 Measurement of vibration levels and spectrum analysis

Vibration levels were measured in $mm\,s^{-1}$ on both non-drive end and drive end bearings using B & K vibration measuring instruments. Table 4.3 shows the vibration levels in horizontal, vertical and axial directions.

4.4.3 Inspection

The dimensional accuracy of new bearings was checked and found to be in line with the specifications. Also, the dimensional accuracy and metallurgical examinations of the components of the failed bearings indicate that the results are in line with the specified norms.

4.4.4 Failed bearings and lubricants

Various failed bearings were examined (refer to Fig. 5.2 in Chapter 5). A sample of fresh grease and used grease from the rolling-elements of the failed bearings were collected and analysed.

4.4.5 Grease pipe contacting the base frame

In a few motors, the outlet grease pipe is either in direct contact with the base

Table 4.1 Overall values of shaft voltages of different motors without grounding brush (in V, rms)

Motor	Non-drive end to drive end shaft	Non-drive end to ground	Drive-end shaft to ground
A	1.004	1.004	1.0
B	0.64	0.030	0.65
C	1.14	1.15	0.47
D	0.85	0.85	0.58
E	0.65	0.80	0.82
F	0.74	0.91	0.40

Table 4.2 Frequency analysis of shaft voltages (V) of different motors

Motor no.	Frequency (Hz)									
	Non-drive end to drive end					Non-drive end to ground				
	50	195	1150	1445	Overall	50	195	1150	1445	Overall
A	0.04		0.10	0.33		0.13	0.03	0.28	0.55	0.74
B	0.67		0.07	0.17	0.81	0.008	0.002	0.009	0.014	0.02
C	0.96	0.15	0.19	0.46	1.44	0.062	0.18	0.31	0.65	0.91
D						0.77	0.1	0.21	0.65	1.22
E	0.18	0.029	0.11	0.37	0.525	0.06	0.026	0.123	0.47	0.53
F	0.87	0.0212	0.053	0.19	0.978	0.55		0.076	0.22	0.76

plate or indirect contact through the accumulation of grease from the bearing (Fig. 5.1).

4.4.6 Grease leakage through seals

Excess grease leaking through seals gets collected just beside the bearing housing and thus the contact between the bearing housing and the base plate is made through the contaminated grease even if the bearing pedestal is insulated from the base plate as shown in Fig. 5.1. The collection of dirt/sludge, etc. between the corners of the base at the bearing pedestal and the base plate was also detected, particularly in thermal powerhouses.

4.4.7 Unshielded instrumentation cables

Instrumentation cable used for measurement of bearing temperature sometimes becomes unshielded for various unforeseen reasons, and cable comes partially in contact with the base plate/bearing housing. A few such cases have been found and investigated. Unshielded instrumentation cables were established as the cause of bearing failure after investigations.

4.4.8 Improper contact of grounding brush with shaft

Owing to prolonged operation, dust collection, an incorrect gap between brush and shaft, and improper maintenance of the grounding brush loosens the grip with the rotating shaft. This was found in various locations.

4.4.9 Damage of bearing insulation

The prolonged operation of bearings under vibration ages the insulation, depending on the quality of the insulating material. This results in breakage/

Table 4.3 Vibration levels of different motors in mm s^{-1} (rms)

Motor	Non-drive end			Drive end		
	Horizontal	Vertical	Axial	Horizontal	Vertical	Axial
A	0.7	1.2	0.7	1.5	1.5	0.8
B	0.7	0.65	0.7	2.0	1.6	1.1
C	0.8	0.8	0.65	1.2	1.1	0.8
D	0.6	0.6	0.55	1.0	0.8	0.7
E	0.65	0.35	0.35	0.8	0.9	0.5
F	0.6	1.0	1.0	2.0	1.5	1.0

As per ISO 10816 Part 1-5, bearings are in good condition up to vibration levels of 1.8 mm s^{-1} and above this are in satisfactory condition.

cracking of the bearing insulation. Consequently a resistance-free path of bearing current is created, which leads to catastrophic failure of the bearings (Fig. 5.1).

4.4.10 Passage of current through connecting bolts, nuts and joints

If the connecting bolts, nuts, joints of bearing pedestal and base plate are not properly shielded, the current may pass through the bearing even if the bearing insulation pads are properly maintained (Fig. 5.1).

4.5 Results and discussion

4.5.1 Shaft voltages and their frequencies

The measurement of shaft voltages with and without a grounding brush indicated that the gap between the shaft and the grounding brush was not set precisely. Also, the brush had not been maintained and cleaned properly, so was unable to ground the shaft voltage adequately. Table 4.1 indicates that differing levels of shaft voltages exist in the motors because of the various causes as discussed. The minimum voltage of motor B between NDE to ground was measured as 0.030 V as against 0.64 V between NDE and DE shaft, and 0.65 V between DE to ground (Table 4.1). This indicates that the insulation at NDE bearing of motor B is bridged on the path. This was confirmed on dismantling the bearings of motor B and breakage of the insulation was detected. The maximum voltage of motor C between NDE and the ground was measured as 1.15 V. The DE shaft to ground voltage of 0.4 V of motor F indicates the passage of feeble current through the bearings compared with motors A, C, D and E. The investigations indicated the partial unshielded instrumentation cable contacting the bearing housing was the source of the passage of current through the bearing of the motor F.

The frequency analysis of the shaft voltage signals shown in Table 4.2 indicates that the major component of shaft voltage consists of the magnitude of the slot passing frequency at 195 Hz, 1150 Hz and 1445 Hz and line passing frequency component of 50 Hz. The voltage of the slot passing frequency component at 1445 Hz between NDE to DE varies between 0.17 and 0.37 V, and between NDE to ground between 0.014 and 0.65 V. The voltage at line frequency component of 50 Hz, between NDE to DE, and NDE to ground varies between 0.04 V and 0.87 V, 0.014 V and 0.55 V, respectively. Figure 4.1 shows the typical voltage frequency spectrum between NDE and DE of motor F.

The different magnitudes at the above frequencies of the shaft voltage were generated because of the magnetic asymmetry at the unequal air gap created by static and dynamic eccentricity. This may be because allowable

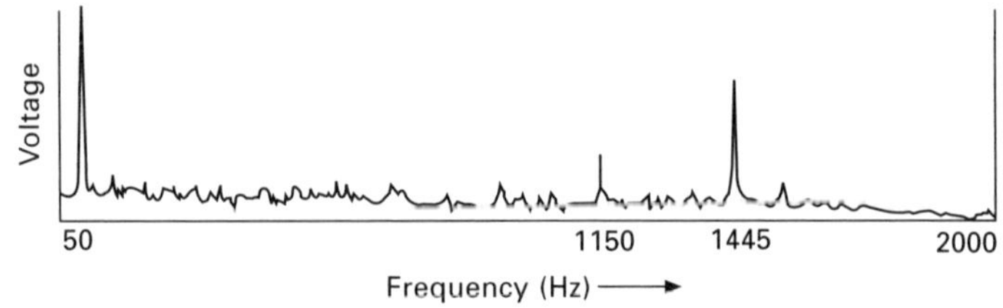

4.1 Frequency spectrum of voltage between NDE and DE (ground) of motor F.

quality control norms were exceeded, and may be controlled to some extent by increasing the precision of the machines.

4.5.2 Vibration analysis

The vibration levels and noise emission indicate the bearing condition. For an unhealthy rolling-element, bearing levels of high-frequency components and overall vibration levels increase considerably. Initially, incipient damage like micro-spalls on the track surface of rolling-element bearings generates high-frequency vibrations. However, when the defects in the roller track increase and grow in size, then the magnitude of low-frequency components increases.[16]

From the magnitude of vibration levels, it is evident that the overall vibration levels in axial and radial directions are within acceptable limits as per ISO 10816, Part 1-5 (Table 4.3). The presence of the principal slot harmonic frequency components and their side bands between 1300 and 1600 Hz exist in all the motor bearings with varying higher levels in the axial direction. This indicates the existence of varying degrees of magnetic asymmetry of air gaps due to static and dynamic eccentricity in all the motors. This, in turn, creates different levels of shaft voltages[1-3] as shown in the Tables 4.1 and 4.2, and leads to bearing failure due to the passage of electric current through bearings brought about by the various unforeseen causes as shown in Section 4.3.

Overall vibration levels in all the motors were within normal limits except in motors B and F, where overall vibration levels were exceeded up to 2 mm s^{-1} as shown in Table 4.3. Moreover, the major constituents of the spectra of motors B and F were found to lie in the low-frequency range, which indicates that the incipient damage had already started in these motor bearings. Shock pulse levels of motors B and F also indicate an initial incipient stage of damage of bearings, whereas the condition of all other motor bearings was normal although they needed immediate relubrication.

4.5.3 Passage of current through bearings

Whenever a grease outlet pipe is either in direct contact with the base plate or in indirect contact through the accumulation of grease from the outlet pipe and the base plate, the circular path of current flow through the rotating shaft and the bearings is closed. This happens even if the bearing insulation and/ or insulation pads are intact. Similarly, the accumulation of grease leaking from bearing seals on the base plate and pedestal reduces the current. Furthermore, accumulation of excessive dirt/sludge between the corner base of the bearing pedestal and base plate creates the path for current to flow even in the presence of insulation pads. This happens when the bearing is not insulated in the housings as shown in Fig. 5.1. If the shaft is not making proper contact with the grounding brush, the brush is not able to make the path of least resistance for grounding the shaft current. The current then tends to pass through the bearing or creates a localized loop in the bearing depending on the bearing impedance. This leads to deteriorating bearing condition and causes bearing failure in due course. This happens when the bearing housing and bearing pedestal are not properly insulated.

Sometimes an unshielded instrumentation cable touching the base plate also creates the path of least resistance for the flow of electric current through the bearings apart from the puncturing of the bearing insulation. This was established as the cause of failure of motor bearings F.

4.5.4 Magnetic flux density

The presence of magnetic flux density along with the corrugation pattern and corrosion on the bearing surfaces indicate the damage due to electric current as explained in references 14 and 15. The presence of a corrugation pattern without significant flux density distribution indicates excessive loading of the bearing surfaces by mechanical means, accompanied by the flexibility of the supporting structure. This is influenced by the frequency of rotation of rolling-elements in the inner race.

4.5.5 Analysis of failed bearings

Corrugation and ridges were found on the track surfaces of the races of all the failed bearings as well as the corrosion on the track surfaces. The track surfaces of the bearings are corroded because of the decomposition of the grease, and formation of corrugations on the track surfaces by low-temperature tempering and Hertzian pressure on the race ways,[1,3] which results in the reduction of bearing life and failure in due course.[8,9]

4.5.6 Effects of bearing current on lubricant

The zinc additive, i.e. zinc dithiophosphate or zinc dialkyldithiophosphate (ZDTP), used as a multifunctional additive in the grease, under rolling friction protects the rubbing metal surfaces and contributes to friction and wear reduction, which depends partly on the amount of additive on these surfaces. Decomposition of ZDTP in the lithium base grease under the influence of electric fields leads to the formation of lithium zinc silicate ($Li_{3.6}Zn_{0.2}SiO_2$) in the presence of a high relative percentage of free lithium and silica impurity in the grease under high temperature in the asperity contacts along with the formation of gamma lithium iron oxide (γ-$LiFeO_2$). During the process, lithium hydroxide is also formed, which corrodes the bearing surfaces. The original structure of lithium stearate changes to lithium palmitate. These changes are not detected under rolling friction. The used grease taken from the motor bearing B showed these changes, similar to that described in references 10 and 11.

4.5.7 Process of bearing failure under the influence of leakage current

When the current leaks through the roller bearing in which low-resistivity ($10^5\,\Omega\,m$) grease has been used, a 'silent' discharge passes through the bearing elements. This creates a magnetic flux density distribution on the bearing surfaces. When a bearing is located and operates under the influence of magnetic field, voltage is generated and current flows through the bearing, depending on the bearing impedance and the threshold voltage phenomenon. Under these conditions, in the initial stages, electrochemical decomposition of the grease occurs, which corrodes the bearing surfaces.[10] Then gradual formation of flutings and corrugations on the surfaces[3,7,13] occurs as found in the bearings of motors B and F. Subsequently, wear increases and the bearing fails.

4.6 Conclusions

From the above investigations and analysis, the following conclusions are drawn:[17,18]

- Current passes through the bearing because of puncturing of the bearing housing insulation, the grease outlet pipe touching the motor base frame and improper contact of the grounding brush with the shaft.
- Current can also pass through a bearing on such occasions when an unshielded instrumentation cable touches the bearing, even if the bearing insulation is healthy. If the bearing is not insulated in the housing, accumulation of dirt/sludge between the pedestal and base plate creates a

path for leakage of current through the bearing even if the pedestal insulation pads are healthy.

- Current passes through a bearing depending on the bearing impedance and threshold voltage phenomenon.
- Levels of measured voltage between bearing and ground and between bearing and shaft indicate the condition of the bearing insulation. Vibration levels indicate the bearing condition.
- In the case of bearings using grease of low resistivity, failure occurs under the 'silent' electric discharge owing to chemical decomposition, formation of flutings and corrugations on the bearing surfaces. Bearings should be properly insulated and shielded for any possible means of leakage of electric current.
- The failure of bearings by the passage of electric current can be established by the detection of the flux density on the bearing surfaces along with the corrugated pattern on the track surfaces, by analysis of deterioration of greases used in the bearings, and by measurement of stray and shaft voltages. The extent of deterioration is ascertained by vibration analysis. Also, the bearings can be shielded for any possible means of leakage of current as reported for their trouble-free operational life.

4.7 References

1. Prashad, H., 'Investigations of Damaged Rolling-element Bearings and Deterioration of Lubricants under the Influence of Electric Current', *Wear*, **176**, 151–161, 1994.
2. Bradford, M., 'Prediction of Bearing Wear due to Shaft Voltage in Electrical Machines', ERA Technology Limited, 1984.
3. Prashad, H., 'Investigations on Corrugated Pattern on the Surface of Roller Bearings Operated under the Influence of Electrical Fields', *Lub. Eng.*, **44**, 8, 710–718, 1988.
4. Prashad, H., 'Effects of Operating Parameters on the Threshold Voltages and Impedance Response of Non-insulated Rolling-element Bearings under the Action of Electric Current', *Wear*, **117**, 223–240, 1987.
5. Prashad, H., 'Magnetic Flux Density Distribution on the Track Surface of Rolling-element Bearings – An Experimental and Theoretical Investigation', *Tribol. Trans.*, **39**, 2, 386–391, 1996.
6. Morgan, A.W. and Whillie, D., 'A Survey of Rolling Bearing Failures', *Proc. Inst. Mech. Eng.* Part F, **184**, 48–56, 1969/70.
7. Andreason, S., 'Effects of an Electric Current on Contact Temperature, Contact Stresses and Slip band initiation on Roller Track of Roller Bearings,' *Ball Bearing J.*, **153**, 6–12, 1968.
8. Simpson, F.E. and Crump J.J., 'Effects of Electric Currents on the Life of Rolling contact Bearings', *Proc. Lubrication and Wear Convention,* Bournemouth, 1963, Institution of Mechanical Engineers, London, 1963, Paper 27, 296–304.
9. Prashad, H., 'Analysis of the Effects of Electric Current on Contact Temperature Contact Stresses and Slip Band Initiation on the Roller Tracks of Roller Bearings', *Wear*, **131**, 1–14, 1989.

10 Prashad, H., 'Diagnosis of Deterioration of Lithium Greases used in Rolling-element Bearings by X-ray Diffractometry', *Tribol. Trans.*, **32** (2) 205–214, 1989.

11 Komatsuzaki, S., 'Bearing Damage by Electrical Wear and its Effects on Deterioration of Lubricating Grease', *Lub. Eng.*, **43**, 25–30, 1987.

12 Remy, M. and Magnin, A., 'Rheological and Physical Studies of Lubricating Greases Before and After Use in Bearings', *Trans. ASME, J. Tribol.*, **118**, 681–686, 1996.

13 Winder L.R. and Wolfe, O.J., 'Valuable Results from Bearing Damage Analysis', *Met. Prog.*, April, pp. 52–59, 1966.

14 Prashad, H., 'Diagnosis of Failure of Rolling-element Bearings of Alternators – A Study', *Wear*, **198**(1–2), 49–51, 1996.

15 Prashad, H., 'The Effect of Current Leakage on Electro-adhesion Forces in Rolling – Friction and Magnetic Flux Density Distribution on the Surface of Rolling-Element Bearing', *Trans. ASME, J. Tribol.*, **110**, 448–455, 1988.

16 Prashad, H., Ghosh, M. and Biswas S., 'Diagnostic Monitoring of Rolling-element Bearing by High-frequency Resource Technique', *ASLE Trans.*, **28**(4) 39–448, 1985.

17 Prashad, H., 'Investigations and Diagnosis of Failure of Rolling-element Bearings Due to Unforeseen Causes – A Case Study', *BHEL J.*, **20**(1), 59–67, 1999.

18 Prashad, H., 'Diagnosis and Cause Analysis of Rolling-element Bearings Failure in Electric Power Equipments Due to Current Passage', *Lub. Eng.*, **5**, 30–35, 1999.

5

Localized electrical current in rolling-element bearings

5.1 A general review

The diagnosis and cause analysis of rolling-element bearing failure have been well studied and established in the literature. Failure of bearings due to unforeseen causes have been reported as: puncturing of bearing insulation; grease deterioration; grease pipe contacting the motor base frame; unshielded instrumentation cable; the bearing operating under the influence of magnetic flux, etc. These causes lead to the passage of electric current through the bearings of motors and alternators, which causes them to deteriorate in due course. Bearing failure due to localized electrical current between track surfaces of races and rolling-elements has not been hitherto diagnosed and analysed.

This study explains the cause of generation of localized current in the presence of shaft voltage. It also brings out the developed theoretical model to determine the value of localized current density depending on dimensional parameters, shaft voltage, contact resistance, frequency of rotation of shaft and rolling-elements of a bearing. Furthermore, failure caused by flow of localized current has been experimentally investigated.

5.2 Introduction

5.2.1 Causes of shaft voltages and flow of current through bearings

Shaft voltages exist in electrical machines as a result of asymmetries of various faults, e.g. winding faults, unbalanced supplies, electrostatic effects, air-gap fields, magnetized shaft and asymmetries of magnetic fields. The causes of shaft voltage can be grouped into four categories:

- external causes;
- magnetic flux in the shaft;

- homo-polar magnetic flux;
- ring magnetic flux.

Furthermore, friction between the belt and pulley can set up an electrostatic voltage between the shaft and bearings. Accidental grounding of a part of the rotor winding to the rotor core can lead to stray currents through the shaft and bearings, and can result in the permanent magnetization of the shaft. Also, a shaft voltage and current could be generated when the machine is rotated. Homo-polar flux can result from an air gap or rotor eccentricity, and this can generate voltage.[1,2]

The most important cause of bearing current is the linkage of alternating flux with the shaft. The flux flows perpendicularly to the axis of the shaft and pulsates in the stator and rotor cores. It is caused by asymmetries in the magnetic circuit of the machine, such as:

- Uneven air gap and rotor eccentricity;
- split stator and rotor core;
- segmented punching;
- axial holes through the cores for ventilation or clamping purposes;
- key ways for maintaining the core stackings; and
- segments of differing permeability.

All the causes listed above result in developing a magnetic flux, which closes in the circumference over the yoke and induces voltage on the shaft as the machine rotates. This results in a localized current at each bearing rather than a potential difference between the shaft ends. A current path, however, along shaft, bearings and frame results in a potential difference between the shaft ends.[3]

It has been reported that at a certain threshold voltage, depending on the resistivity of the lubricant and operating conditions, current flows through the bearing.[4] Also, it was established by Busse *et al.*[5] that the dielectric strength of the lubricant is the ability to withstand voltage without breakdown. Prashad[6] found that the flow of circular current in the inner race leaks through the rolling-elements to the outer race by following a path of least resistance and establishes a field strength, leading to the development of magnetic flux on the track surface of races and rolling-elements. This has been experimentally investigated.

Studies were reported on the causes and control of electrical currents in bearings by Morgan and Wyllie.[7] Andreason[8] determined the flow of current through lubricated contacts. Simpson and Crump[9] studied the effect of electrical current on bearing life. The effect of operating parameters on an impedance response of a rolling element bearing has been analysed by Prashad.[4,10] The deterioration of lubricants used in non-insulated bearings investigated by Komatsuzaki,[11] Prashad[12] and Remy and Magnin.[13] Investigations were also carried out to evaluate the pitting mechanism on the lubricated surfaces under an AC electric current by Lin *et al.*[14] and Chiou *et al.*[15]

The surveys of the failure of rolling-element bearings indicate various causes, including that of failure due to corrugations by Winder and Wolfe[16] and Prashad.[17,18] The mechanism of corrugations formed on the track surfaces and related investigations have been reported by the author.[3,19] In addition to the well-established causes, the bearing may also fail due to causes that are generally unforeseen during design and operation. These causes have been discussed by Prashad.[20]

This chapter presents, with the help of a case study, the failure of bearings of motors by the effect of a localized electrical current between the track surface of races and rolling-elements under the conditions when the other causes of bearing failure have been ruled out by all possible investigations. The principle for the generation of a localized current is established and analysed along with the development of theoretical model to determine its value.

5.3 Bearing arrangement and the nature of bearing failure

Failure of bearings was detected in a few motors. These motors, rated 2100 kW and operating at 1494 rpm, have a three-bearing arrangement. The motors are used to drive primary air fans.

At the non-drive end (NDE) of the motor, bearing type NU 228 is used and is insulated. Figure 5.1 shows the arrangement of the bearing in the motor. At the drive-end (DE) of the motor, roller bearing type NU 232 and ball bearing type 6332 are used for taking both radial and axial loads simultaneously. All the bearings are grease-lubricated, and the motors are designed for continuous operation with periodic relubrication. Thermocouples were provided for continuous measurement of the temperature during operation.

After commissioning, a few motor bearings on both NDE and DE failed prematurely. The nature of the failure was due to the formation of corrugations and flutings on the roller tracks of races besides corrosion on the raceways and rolling-elements, irrespective of the make of the bearings used. A typical corrugation pattern on the track surface of a failed bearing is shown in Fig. 5.2. However, the depth of corrugations and degree of development of corrugations and flux density were different on various failed bearings. The colour of the grease between the rolling-elements was also found to change, becoming black.

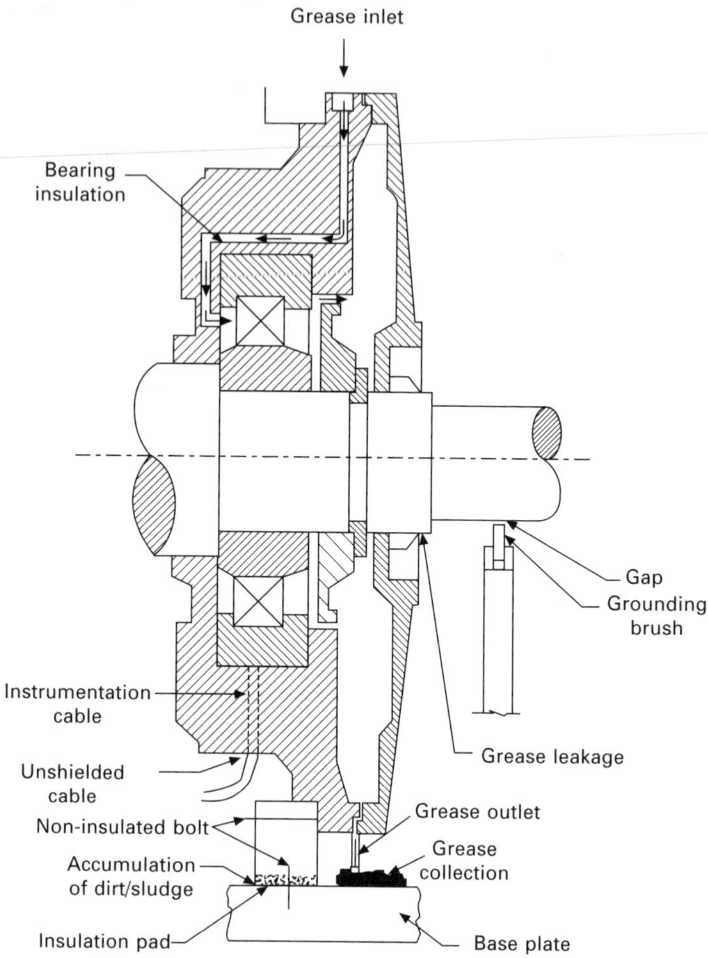

5.1 Schematic showing bearing arrangement with lubricant flow path, unshielded instrumentation cable and other unforeseen causes leading to the passage of electric current.

5.4 Investigations, observations and data collection

Various investigations pertaining to shaft voltage and vibrations were carried out on bearings of different motors, both on NDE and DE, and various other related aspects were inspected. The shaft voltages were measured between NDE and DE shaft, DE shaft-to-ground, and NDE shaft-to-ground. Vibration levels were measured in micrometres and mm s^{-1} at both NDE and DE bearings. The dimensional accuracy of new bearings were checked and found to be in line with the specifications. Inspection of dimensional accuracy and

5.2 Typical corrugation pattern on the track surface of a failed bearing under the influence of electrical current.

metallurgical examinations of the components of the failed bearings also revealed consistency with the specifications. Various failed bearings were examined. Samples of fresh grease and used grease from the rolling-elements of the failed bearings were collected and analysed.

Potential causes of bearing failure, which might have led to passage of current though bearings, were ruled out as follows (Fig. 5.1):

- Grease pipe was not in contact with the base frame in any of the motors.
- Current path through bearings via collection of leaked grease and base frame was not detected.
- All the instrumentation cables werc completely shielded.
- Insulation of the bearings was fully intact in all the failed motor bearings. This was established by value of high resistance measured across the bearing insulation.
- Passage of current through connecting bolts, nuts, etc. could not be established in any of the failed motor bearings.

The shaft voltage at the site on these failed motor bearings was measured as 1600 mV before and after the replacement of failed bearings. The shaft voltage at the works during the test run was as high as 500 mV for these motors. The difference of shaft voltage measured at the sites and the works may be attributed to the analogue/digital devices used as well as the coupled mode of operation and other unforeseen causes. For other normal running motors, shaft voltage as measured was less than 200 mV. The failure of NU 228 bearings was reported at the site on these motors after a run of approximately 300 h. A grounding brush was not used in all such motors. Furthermore, there was not adequate space available on the shaft to install the grounding brush.

5.5 Theoretical model and approach to determine the flow of localized current in a bearing

In a rolling-element bearing using low-resistivity lubricant ($10^5\,\Omega\,m$) in the presence of shaft voltage, current can flow through the bearings. This occurs in all such cases when current finds a low-resistance path to flow as discussed in Section 5.4. However, when the current does not find a way to pass through any of the paths as listed in Section 5.4, and the shaft voltage is more than the threshold voltage/safe voltage (>200 mV), a substantial level of localized current between the rolling-elements and the track surface of the inner race may appear which may cause the bearing to deteriorate and the lubricant to disintegrate with the passage of time, causing flux density distribution, corrosion and blackening of the track surface as reported by Prashad.[20] This reduces the fatigue life of bearings and leads to premature failure.

The existing shaft voltage on the rotating shaft induces a voltage on the track surface of rolling-elements. Since the rolling-element rotates at a much higher frequency than the shaft rotating speed[19,21] this generates a higher voltage on the surface of rolling-elements as per the principles of electric fields and electromagnetism.[22] Thus, a potential difference between the inner race and rolling-elements develops. Revolution of rolling-elements along the track surface further affects the induced voltage on the track surface of the rolling-elements as well as their rotation around their own axis. This makes the analytical model quite complex. However, the potential difference between rotating shaft/inner race and rolling-elements leads to the passage of localized current at the contact surface, depending on the contact resistance between the contact zone of track surface of the inner race and the rolling-elements. The induced voltage phenomenon also occurs between the track surface of the outer race and the rolling-elements. Furthermore, the flow of localized current between the inner race and the rolling-elements, depending on the different points of contacts during rotation, turns into a circular current in the inner race and rolling-elements quite frequently in the course of operation. The proximity contact of rolling-elements with the outer race may lead to a flow of current through the outer race for a short duration. This flow of partial/full circular current establishes the field strength, leading to the development of residual magnetic flux on the track surface of races and rolling-elements in due course.[6]

It may be imagined that current enters the track surface of the inner race because of the developed potential drop between rolling-elements and track surface of the inner race in the loaded zone at the surfaces of contact points between them, in a distributed form. The current flows around and outward until it concentrates at one or several rolling-contacts. As the rolling-elements orbit, the current carrying elements travel. Current then flows through each of the conducting rolling-elements, between two diametrically opposite contact areas. As the rolling-elements rotate with respect to the contact areas, these currents gradually might sweep out 360° around a major circle. If there was only one conducting rolling-element at a time, then two arc currents would travel in the race towards its contact areas, one clockwise and one counter-clockwise. Depending on the resistance, these two currents may or may not be equal in magnitude. If more than one rolling-element contacts, then, of course, the situation is more complex. In the theoretical model, current flow in one direction in the inner race is assumed for analysis. The flow and break of flow of localized electrical current continue to depend on condition and regime of operation.

5.5.1 Determination of speed of rolling-elements

Under pure rolling, the absolute velocity at a point located on the circumference of a rolling-element is equal to the circumferential speed of the inner race. Simultaneously, an opposite point (at 180°) of the same rolling-element contacts the stationary outer race. The absolute speed of each point on the rolling-element is a combination of the speed of the rolling-element around its axis and the speed of the set of rolling-elements around the bearing axis. The absolute velocities of the two points 180° apart on the diameter of the rolling-element are parallel and vary linearly between the velocities at the points of contact on the inner and outer races.

With the above analysis, the rotating speed frequency of the rolling-elements has been determined as:[19]

$$f_b = 0.5\, f_s(D/d)\,[1 - (d/D)^2 \cos^2\alpha] \qquad [5.1]$$

The ratio of the speed of rotation of rolling-elements and shaft speed is given as:

$$V_{bs} = f_b/f_s = 0.5\ D/d[1 - (d/D)^2 \cos^2\alpha] \qquad [5.2]$$

For a radial cylindrical roller bearing, $\alpha = 0$. So:

$$V_{bs} = 0.5[(D^2 - d^2)/dD] \qquad [5.3]$$

5.5.2 Induced voltage on rolling-elements

As the shaft voltage E develops and changes with time, the voltage on the track surface of the inner race E_{ir} changes accordingly. This is because the inner race is press fitted on the shaft and is considered integral with the shaft. The voltage E_{ir} induces the voltage on the rolling-elements rotating around their own axis and moving along the track surface of the inner and outer races. This makes the phenomenon of induced voltage very complex.

The rotating rolling-elements loop may be considered as a type of AC generator or, in other words, rolling-elements develop voltage depending on the speed of rotation. The voltage in absolute terms is more on those surfaces of the rolling-elements, which are at right angles to the existing field, because of the shaft voltage on the track surface of the inner race and is zero on the parallel surfaces to the field. The developed voltage on rolling-elements broadly depends on the area of such a loop formed by rotating rolling-elements and not on their shapes. The voltages thus developed on the surfaces of rolling-elements E_{ir} depend on the rotating speed of the rolling-elements as well as other electromagnetic factors. Considering the ratio of the speed of rolling-elements with respect to the shaft speed V_{bs} the voltage on the rolling-elements E_r is taken as the product of V_{bs} and the shaft voltage E using principles of electrical current theory.[22] This leads us to determine the potential difference between rolling-elements and track surface of inner race as:

$$E_{rir} = E_r - E_{ir} \qquad [5.4]$$

On using Equation (5.3), the potential difference E_{rir} is determined as

$$E_{rir} = E\, V_{bs} - E \qquad [5.5]$$

where $E_{ir} = E$ and $E_r = E\, V_{bs}$

or,

$$E_{rir} = E\left[\frac{(D^2 - d^2 - 2Dd)}{2Dd}\right] \qquad [5.6]$$

Furthermore, development of the same induced voltage as on rolling-elements occurs on the track surface of the outer race under the influence of voltage existing on the surfaces of rolling-elements. Thus, the potential difference between the rolling-elements and the outer race will be insignificant as the outer race of the bearing is stationary.[22]

5.5.3 Bearing resistance

The resistance between the track surface of the inner race and the rolling-elements depends on bearing kinematics, lubricant film thickness and width of contact between track surface and rolling-elements and the number of rolling-elements in loaded zone. For the bearing using low-resistivity lubricant ($10^5\,\Omega$m), the resistance (R_c) of the bearing as determined is found to vary between 0.1 and 0.5 Ω.[4]

5.5.4 Localized electrical current in bearings

Localized current between the track surface of the inner race and the rolling-elements is determined using Equation (5.6) and the contact resistance (R_c), and is given as:

$$I_b = \frac{E_{rir}}{R_c} = \frac{E}{R_c}\left[\frac{D^2 - d^2 - 2Dd}{2Dd}\right] \quad [5.7]$$

5.6 Field strength on track surfaces of races and rolling-elements

Because of the presence of localized electrical current in the partial/complete arc of the inner race and rolling-elements, the field strength develops on the track surfaces of rolling-elements and the inner race in due course. The field strength on these elements can be determined separately as reported by Prashad.[6]

5.6.1 Field strength on the track surface of inner race

The field strength H_{irr} on the track surface of the inner race due to flow of current in rolling-elements has been analysed, and is determined as:[6]

$$H_{irr} = 2\pi I_b[R^2 - R_{ir}^2]/R^3 \quad [5.8]$$

5.6.2 Field strength on rolling-elements

On rotation, rolling-elements change polarity and, the outer race being stationary, the field strength on the surface of rolling-elements depends on the flow of circular current in the inner race and is given as:[6]

$$H_{rir} = 2\pi I_b R_{ir}(R^2 - R_{ir}^2)/R^4 \quad [5.9]$$

The chances of development of significant field strength on the track surface of the outer race due to the flow of intermittent local current in the track arcs of inner race and rolling-elements are quite remote, and hence it is not considered in the analysis. However, sometimes development of traces of flux density on the outer race track surface cannot be ruled out.

5.7 Magnetic flux density

The field strength on the track surface of the inner race and rolling-elements of the bearing gives rise to the development of magnetic flux density on these surfaces. These can be determined analytically as reported by Prashad.[6]

5.7.1 Magnetic flux density on the track surface of the inner race

The magnetic flux density on the track surface of the inner race of the bearing B_{ir} due to field strength H_{irr} in an oil medium of relative permeability U_r, with respect to free space is given (in tesla) by Prashad:[6]

$$B_{ir} = U_o U_r H_{ir} = 4\pi \times 10^{-7} U_r H_{irr} \qquad [5.10]$$

which is determined (in gauss) using Equation (5.8) as:

$$B_{ir} = 78.96 \times 10^{-3} U_r I_b (R^2 - R_{ir}^2) / R^3 \qquad [5.11]$$

5.7.2 Magnetic flux density on the track surface of rolling-elements

On using Equations (5.9) and (5.10), the residual flux density on a few rolling-elements (in gauss) is determined as:

$$B_r = 78.96 \times 10^{-3} U_r I_b R_{ir} (\mathrm{R}^2 - R_{ir}^2)/R^4 \qquad [5.12]$$

In may be noted that the residual magnetization of steel, which remains after an alternating magnetizing current is switched off, bears no simple theoretical relationship to the magnetization during passage of a direct current of the same effective value.

5.8 Determination of time span for the appearance of flutes on the track surfaces

Instant thermal stresses due to thermal transients on the roller track of races caused by roller contact under the influence of electrical current depend on an instant rise in temperature. As the temperature rise stabilizes, the contact thermal stresses increase and affect the fatigue life. The duration the bearing would have taken after the formation of slip bands and before the appearance of flutes/corrugations on the track surface of the inner race is determined as:[23]

$$t = \frac{(\pi \partial_{corr})^2 (BD_{ir} + BD_{or} + LdN)(BD_{ir}\,\Delta_{ir}^{-1} + BD_{or}\,\Delta_{or}^{-1} + LdN\,\Delta_r^{-1})}{2E_{rir} I_b Y}$$

5.9 Data deduction

All the failed bearings both on NDE and DE ends of the three motors were examined. In these bearings, the damage as reported in Section 5.3 was studied. Investigations were carried out on NU 228 bearings and flux density distribution on their track surfaces was detected using a Hall probe, similar to the procedure reported by the author.[6] The pitch of corrugations on the track surfaces of the NU 228 bearing was measured. The theoretical time span for the formation of corrugations after the slip band formation was determined using Equation (5.13). The dimensional and operating parameters

Table 5.1 Dimensional and operating parameters of NU 228 bearing

B	=	42 mm
R_{ir}	=	79 mm
R_{oro}	=	125 mm
R_{or}	=	116 mm
R	=	97.5 mm
d	=	37 mm
D	=	195 mm
D_{ir}	=	158 mm
D_{or}	=	232 mm
L/B	=	1
N	=	14
n	=	1500 rev min^{-1}

Table 5.2 Experimental and theoretical data

E	=	1.6 V
R_c	=	0.20 Ω
E_{rir}	=	2.46 V
E_{rir}/E	=	1.54
I_b	=	12.30 A
Y	=	210×10^9 N m^{-2}
B_{ir}	=	3.6 G (theoretical)
B_{ir}	=	5 G (maximum on measurement)
B_r	=	3 G (experimental on a few rolling-elements)
B_r	=	2.7 G (theoretical)
Δ_{ir}	=	0.25 mm
Δ_{or}	=	not detected
Δ_r	=	not detected
∂_{corr}	=	700×10^6 N m^{-2}
t	=	107.45 h
t_1	=	300 h

of bearing type NU 228 and various values of measured/analytical parameters are given in Tables 5.1 and 5.2, respectively.

5.10 Results and discussion

5.10.1 Potential difference between rolling-elements and track surface of the inner race and flow of localized current between them

The potential difference between the rolling-elements and the track surface of the inner race of an insulated bearing working under the influence of shaft voltage depends on shaft voltage, pitch diameter and diameter of rolling-elements as per the derived relations in Equation (5.6). Bearing type NU 228, working under the influence of shaft voltage of 1.6 V, develops a potential difference of 2.46 V between the rolling-elements and the track surface of the inner race. The ratio of this potential difference to the shaft voltage E_{rir}/E_r has been worked out as 1.54 (Table 5.2). This ratio depends on the speed of rotation of rolling-elements to the shaft speed and is a function of dimensional parameters of a roller bearing, Equation (5.3). Furthermore, localized electric current between the track surface of the inner race and rolling-elements depends on the potential difference E_{rir} and contact resistance R_c. It is a function of shaft voltage E contact resistance R_c and dimensional parameters of the bearing Equation (5.7).

The intensity of local current has been determined theoretically as 12.30 A for the existing shaft voltage of 1.6 V and resistance R_c of 0.2 Ω. (Table 5.2). The localized electrical current damages the track surfaces of the bearing in due course and leads to form corrugations. In contrast, for shaft voltages of 500 and 200 mV, the value of local current has been determined as 3.84 and 1.54 A, respectively. These low values of current do not affect the bearing during operation. In addition, the outer race being stationary, no significant potential difference is generated between the outer race and the rolling-elements, and thus current does not pass between them. However, a momentary circular current, through an arc of the outer race, can flow as and when proximity contact of the outer race and rolling-elements takes place in a sector owing to instabilities/vibrations during operation.

5.10.2 Residual flux density distribution on the track surface of inner race and rolling-elements

A flow of current between the track surface of inner race and rolling-elements due to the existing potential drop leads to the generation of a residual flux density distribution on the track surface of inner race and rolling-elements in due course. The flux density on the track surface of the inner race has been determined theoretically as 3.6G, using Equation (5.8). The maximum value of flux density was determined as 5 G by measurement (Table 5.1). On a few rolling-elements flux density was determined and found to be a maximum of 4G. Theoretically, flux density was determined as 2.7 G on rolling-elements using Equation (5.9). The presence of flux density by measurement and theoretical investigations indicates that the localized current passed through the bearing arc partially.[6] This created corrugations/flutes on inner race track surface of the bearings and deteriorated the lubricant, which has corroded the surfaces of bearings and led to failure by the reduction of fatigue life.[10,12,14]

5.10.3 Assessment of time span before appearance of flutes on the track surfaces

As a result of the shearing of atomic planes within the crystals, some crystals on the roller track of the inner or outer race develop slip bands. The slip bands are formed in the subsurface prior to the appearance of flutes/corrugations on the track surfaces. The formation of slip bands is initiated by shear stress caused by operating parameters, and is accelerated by the passage of electric current, corrosion and oxidation of track surfaces. Furthermore, formation of slip bands depends on lubricant characteristics and quality of a bearing. The following may take place before the initiation of corrugations:

- Generation of persistent slip bands (PSB).
- Crack form along PSB and initiated from the tip of slip bands.
- Formation and propagation of flutes/corrugations on the track surface.

After the cracks/slip bands under the track surfaces are initiated, the process that governs the propagation leading to the formation of corrugations/flutes is considered by the continuum theory of Griffith.[10,23] The time a bearing takes before formation of flutes on the track surfaces after the slip bands formation depends on the dimensions of the track surface, pitch of corrugations, number of rolling-elements, potential difference between the track surface and rolling-elements, intensity of local current and properties of the bearing material. The pitch of corrugations on rolling-elements and outer race was not found and hence the values of Δ_r^{-1}, Δ_{or}^{-1} are negligible, Equation (5.13). This time span t for the initiation of corrugations is determined as 107.45 h, using Equation (5.13), without considering the time for formation of slip bands (Table 5.2). The net time for the development of slip bands, including that of flute formation on track surface of the inner race after commissioning, is found to be approximately 300 h according to site data as given in Table 5.2. This might match practically with the actual time of initiation of flutes after formation of slip bands, as determined analytically.

5.10.4 Bearing failure under localized current

Electrical current damage of a bearing is of two types. In the first type, when the low-resistivity lubricant ($< 10^5\,\Omega\,\text{m}$) is used in the roller bearings, a silent discharge occurs through the bearing elements under the influence of electrical current. This breaks down the used lubricant and corrodes the surfaces of the bearings, which lowers the fatigue life of the bearings as discussed by Prashad.[3,12] Furthermore, the passage of current through a bearing increases the bearing's operating temperature. Subsequently, thermal stresses are increased and surface heating takes place, which leads to low-temperature tempering of the track surfaces. This accelerates the formation of slip bands and corrugations on the track surfaces in due course.[10,18] Magnetic flux density is also developed on the track surfaces, and the original structure of the lubricant undergoes changes.[6,12]

In the second type of bearing failure, when the high-resistivity lubricant ($> 10^9\,\Omega\,m$) is used in the roller bearings, an accumulation of charges occurs on the track surfaces until it reaches a threshold critical value, when breakdown takes place. This leads to damage of the track surface caused by arcing. This is accompanied by mass transfer and elevated local temperature on the asperity of the contact surfaces.[4,25]

5.11 Conclusions

Based on the above analysis, the following conclusions are drawn:[26]

- Under the influence of higher shaft voltage, a rolling-element bearing using low-resistivity lubricant may deteriorate because of the development of a potential difference between the track surface of the inner race and rolling-elements. This leads to the passage of localized current, depending on the contact resistance between them.
- The potential difference between the track surface of the inner race and rolling-elements develops because of the higher frequency of rotation of rolling-elements compared with the shaft speed. The potential difference, thus developed, depends on shaft voltage and bearing kinematics.
- The ratio of potential difference between track surfaces of inner race and rolling-elements to that of shaft voltage is a function of pitch and rolling-elements diameter of a bearing. For the NU 228 bearing, it is determined as 1.54.
- The development of magnetic flux density on the track surface of the inner race and rolling-elements indicates flow of locally generated current between them.
- The time of appearance of flutes on the track surface can be estimated by bearing kinematics, existing potential difference between track surface of inner race and rolling-elements, value of localized current, properties of bearing material, together with measured values of pitch of the corrugations on the track surfaces.
- The failure of a bearing by the localized current can be avoided by limiting the shaft voltage to a maximum of 200 mV. If a higher value of shaft voltage persists, it is preferable to ground a brush. However, altering the path of flow of circular current, through dismantling and reassembling of the bearing as a means of modulating the track surfaces under interaction, the effect of localized current can sometimes be minimized.
- Localized current in rolling-element bearings is a complex phenomenon and needs the solution of complex analytical models and investigations.

5.12 References

1 Prashad, H., 'Investigations of Damaged Rolling-element Bearings and Deterioration of Lubricants under the Influence of Electric Current,' *Wear*, **176**, 151–161, 1994.
2 Bradford, M., 'Prediction of Bearing Wear due to shaft Voltage in Electrical Machines', *Technical Report No. 84–007*, ERA Technology Limited, England, pp. 49–53, 1984.
3 Prashad, H., 'Investigations on Corrugated Pattern on the Surface of Roller Bearings Operated Under the Influence of Electrical Fields', *Lubr. Eng.*, **44**(8), 710–718, 1988.
4 Prashad, H., 'Effects of Operating Parameters on the Threshold Voltages and Impedance Response of Non-insulated Rolling-element Bearings under the Action of Electric Current', *Wear*, **117**, 223–240, 1987.
5 Busse, D., Erdman, J., Kerkman, R.J., Schlegel, D. and Skibinski, G., 'System Electrical Parameters and their Effects on Bearing Currents', *IEEE Trans. Ind. Appl.* **33**(2) 577–584, 1997.
6 Prashad, H., 'Determination of Magnetic Flux Density on the Surfaces of Rolling-element Bearings as an Indication of the Current that has Passed Through Them – An Investigation', *Tribol. Int.*, **32**, 455–467, 1999.
7 Morgan, A.W. and Wyllie, D., 'A Survey of Rolling Bearing Failures', *Proc. Inst. Mech. Eng., Part F* **184**, 48–56, 1969–70.
8 Andreason, S., 'Passage of Electric Current through Rolling Bearings', *Ball Bearing J.* **153**, 6–12, 1968.
9 Simpson, F.E. and Crump, W.J.J., 'Effects of Electric Currents on the Life of Rolling Contact Bearings', *Proc. Lubrication and Wear Convention*, Bournemouth, Inst. Mech. Eng., London, Paper 27, pp. 296–304, 1963.
10 Prashad, H., 'Analysis of the Effects of Electric Current on Contact Temperature Contact Stresses and Slip Band Initiation on the Roller Tracks of Roller Bearings', *Wear*, **131**, 1–14, 1989.
11 Komatsuzaki, S., 'Bearing Damage by Electrical Wear and its Effects on Deterioration of Lubricating Grease', *Lubr. Eng.*, **43**(1), 25–30, 1987.
12 Prashad, H., 'Diagnosis of Deterioration of Lithium Greases used in Rolling-element Bearings by X-ray Diffractometry', *Tribol. Trans.*, **32** (2), 205–214, 1989.
13 Remy, M. and Magnin, A., 'Rheological and Physical Studies of Lubricating Greases Before and After Use in Bearings', *Trans. ASME, J. Tribol.*, **118**, 681–686, 1996.
14 Lin, C.M., Chiou, Y.C. and Lee, R.T., 'Pitting Mechanism on Lubricated Surface of Babbit Alloy/Bearing Steel Pair under AC Electric Field', *Wear*, **249**, 133–142, 2001.
15 Chiou, Y.C., Lee, R.T. and Lin, C.M., 'Formation Criterion and Mechanism of Electrical Pitting on the Lubricated Surface under AC Electrical Field', *Wear*, **236**, 62–72, 1999.
16 Winder, L.R. and Wolfe, O.J., 'Valuable Results from Bearing Damage Analysis', *Metal Progress*, **4**, 52–59, 1968.
17 Prashad, H., 'Diagnosis of Failure of Rolling-element Bearings of Alternators – A Study', *Wear*, **198**, 46–51, 1996.
18 Prashad, H., 'The Effect of Current Leakage on Electro-adhesion Forces in Rolling – Friction and Magnetic Flux Density Distribution on the Surface of Rolling-element Bearing', *Trans. ASME, J. Tribol.*, **110**, 448–455, 1998.
19 Prashad, H., Ghosh, M. and Biswas, S., 'Diagnostic Monitoring of Rolling-element Bearing by High-frequency Resource Technique', *ASLE Trans.*, **28**(4), 439–448, 1985.

20 Prashad, H., 'Diagnosis and Cause Analysis of Rolling-element Bearing Failure in Electrical Power Equipment Due to Current Leakage', *Lubr. Eng.*, **55**(5), 30–35, 1999.

21 Prashad, H., 'The Effect of Cage and Roller Slip on the Measured Defect Frequency Reference of Rolling-element Bearings', *ASLE Trans*, **30**(3), 360–66, 1987.

22 Starling, S.G., *Electricity and Magnetism*, 7th ed., Longmans, Green and Co., New York, 1960.

23 Prashad, H., 'Determination of Time Span for Appearance of Flutes on Track Surface of Rolling-element Bearings Under the Influence of Electrical Current', *Tribol. Trans*, **41**(1), 103–109, 1998.

24 Prashad, H. and Murthy, T.S.R., 'Behavior of Greases in Statically Bounded Conditions and When Used in Non-insulated Anti-friction Bearings Under the Influence of Electrical Fields', *Lubr. Eng.*, **44**(3), pp. 239–246, 1988.

25 Prashad, H., 'Theoretical Analysis of the Effects of Instantaneous Charge Leakage on Roller Track of Roller Bearings Lubricated with High Resistivity Lubricants', *Trans. ASME J. Tribol.*, **112**, 37–43, 1990.

26 Prashad, H., 'Diagnosis of Rolling-element Bearings Failure by Localized Current Between Track Surfaces of Races and Rolling-Elements', *Trans. ASME J. Tribol.*, **124**, 468–473, 2002.

5.13 Nomenclature

B width of track surface
B_{ir} magnetic flux density on track surface of inner race
B_{or} magnetic flux density on track surface of outer race
B_r magnetic flux density on rolling-elements
d diameter of rolling-element
D pitch diameter
D_{ir} outside diameter of inner race
D_{or} inner diameter of outer race
E shaft voltage
E_r voltage on rolling-elements
E_{ir} voltage on track surface of rotating inner race
E_{rir} potential difference between rolling-elements and track surface of inner race
f_s shaft rotational frequency
f_b rolling-element frequency
H_{irr} field strength on track surface of inner race due to flow of current in rolling-elements
H_{rir} field strength on surface of rolling-elements due to flow of current in inner race
I_b bearing current
L length of rolling-element
N number of rolling-elements in a bearing
n rpm

R_c	bearing resistance
R_{ir}	track radius of inner race
R_{or}	track radius of outer race
R_{oro}	outside radius of outer race
R	pitch radius of bearing
t	time required for appearance of flutes after slip band formation on track surfaces
t_1	net time of bearing operation after commissioning for inspection
U_r	relative permeability oil with respect to free space (=1)
U_o	permeability of free space (4×10^{-7} Hm^{-1})
V_{bs}	ratio of speed of rolling-elements and shaft speed (f_b/f_s)
Y	Young's modulus of elasticity
$\propto$	contact angle
∂_{corr}	stress for flute appearance from opening of the tip of flute from an existing slip band on the track surface
Δ_{ir}, Δ_{or}, Δ_r	pitch of corrugation on inner race, outer race and rolling-element, respectively

6 Response and performance of a rolling-element bearing under the influence of an electric current

6.1 A general review

In this chapter the response and performance of roller bearings operating under the influence of an electrical current are analysed and the effect of electrical current on the lubricating greases, analysis of pitch and width of corrugations formed on the roller track of races, threshold voltage phenomenon, impedance response and electro-adhesion forces in rolling friction are discussed. Also, the roles of bearing kinematics and operating conditions on the capacitance, impedance and charge accumulation are discussed. The methodology to determine contact stresses, rise in contact temperature and the number of cycles before the slip band/crater initiation on the track surfaces is also assessed. The effects of the capacitive response of bearings on repeated starts and stops of a machine and instantaneous temperature rise owing to discharge of the accumulated charges are highlighted. This chapter gives an analysis of various aspects of roller bearings working under the influence of different levels of shaft voltages and a discussion of the mechanism of bearing failure and makes it possible to predict the life of a roller bearing lubricated with lubricants of different resistivities. It also gives the state-of-the-art of bearing response and provides the potential to analyse the performance of a roller bearing operating under the influence of electric currents.

6.2 Introduction

Various surveys have indicated that about 30% of all motor failures are due to bearing damage accounted for by the bearing current. Insulating the bearings from the shaft and machine frame can prevent this. Such insulation is an additional complication in machine design and construction, and is difficult to incorporate in flameproof enclosures. Intermittent shorting of this insulation might occur inadvertently in service, leading to sparking, which is dangerous in a hazardous environment. This also re-establishes the bearing current and tends to make the bearings deteriorate. The effect of electrical currents on bearings in different modes of operation needs to be investigated thoroughly.

Magnetic flux develops in electrical machines due to dissymmetry of the magnetic circuits, which close in the circumference over the yoke, and induces a voltage on the shaft. Shaft voltages can be established in rotating machines of any type, but more particularly in electrical machines. The causes of such voltages can be categorized as external causes, e.g. electrostatic effects, operational faults, magnetic flux in the shaft, homopolar magnetic flux and ring magnetic flux. In general, shaft voltages exist in electrical machines as a result of asymmetry of faults, i.e. winding faults, unbalanced supplies, air-gap fields, magnetized shaft or other machine members, asymmetries of magnetic fields, etc. Asymmetries are caused by rotor eccentricity, poor alignment, manufacturing tolerances, uneven air gaps, segmental lamination punching, variation in permeability and various other unforeseen reasons. These asymmetries in electrical machines result in a net flux or a current linking with the circuit consisting of shaft, bearing and frame. Shaft flux results in localized currents at each bearing rather than a potential difference between shaft ends. Furthermore, under the influence of a potential drop across a roller bearing, the varying film thickness between races and the rollers form a capacitor of varying capacitance, depending on the permittivity of the lubricant. The minimum film thickness between the races and the rollers offers maximum capacitance and minimum capacitive reactance. The active resistance offered by a bearing is minimum at the minimum film thickness; however, it is primarily governed by the resistivity of the lubricant. Under the influence of the shaft voltage, the electrical interaction between the races and the rollers in the presence of the oil film is like a resistor–capacitor (RC) circuit and offers impedance to the current flow.

Fatigue and surface distress usually describe the limits for reliable operation of a rolling-element bearing. Fatigue initiating in the subsurface is better understood than the fatigue initiated on the surface itself, i.e. surface distress. In general, analysis of bearing failures seldom indicates subsurface-initiated fatigue as the reason for failure. The most common form of bearing failure is caused by surface distress. This includes different types of damage including micro-pitting, smearing, indentations and plastic deformation as well as surface corrosion. Obviously, as per the application demands, bearings work under higher temperatures and heavily stressed conditions. Such conditions are unavoidable when bearings are to operate under the influence of shaft voltage and other rigorous operating conditions. This leads to technical innovation covering materials, modified bearing surface coatings and lubricants. Furthermore, this opens a new phase for the evaluation of bearings. In addition, under the effect of a shaft voltage, dirt, metallic particles and irregular lubricant film permit the lubricant oil film to be pierced by the electric current. Under these conditions, the impedance of the bearing circuit becomes so low that small shaft voltages may cause substantial bearing currents. If currents are reduced by using high-resistivity lubricant or by establishing a non-conducting oil film in the bearing, the self-inductance of the single loop of the shaft, bearing and casing may then cause a relatively high induced voltage across the oil film. If this induced voltage is exceeded, the threshold voltage may again break down the lubricating film. Until then, when the voltage is below the threshold voltage, the bearing gives the capacitive response and stores the charges between the rolling elements and the roller track of races.

Much of the work referred to above has been carried out and the purpose of this chapter is to highlight the recent investigations on response and performance of bearings under the influence of shaft voltages. The literature on this subject is very scarce and efforts are being made to understand the phenomenon. It is often asked whether the bearing damage is due to leakage of electrical current. In general, when the bearings carry current as a necessary part of an electrical circuit, the current is self-induced as a result of the design characteristics of the machine, or the current is due to an electrostatic phenomenon, the bearing current affects the lubricating media (grease, oil), bearing surfaces and response characteristics of the bearings. The investigations carried out on lubricants and bearing surfaces indicate in-depth behaviour of bearings under the influence of an electrical current, which has been highlighted in the following sections.

6.3 Behaviour of grease in non-insulated bearings

Recent investigations indicate that the resistivity of a grease depends on its EP properties, viscosity, torque characteristics and consistency. The resistivity of grease also varies with respect to voltage and time. The difference in resistivities among greases can be as high as 10^5 times. The change in resistivity depends on the nature of the impurities or by-products and the types of additive in the greases, besides their density, compressibility and structure. Low-resistivity grease ($10^5\,\Omega\,m$) tends to 'recoup' its resistivity when the applied field is switched off. The percentage of 'recouping' varies from 18 to 82 and depends on the stretching of the molecules. Under the influence of an electric field, the carboxylic group at $2660\,cm^{-1}$ of the low-resistivity lithium-base grease decomposes and the lithium metal concentration in the aqueous solution increases relatively. The decomposition of carboxylic acid leads to corrosion of the bearing track surfaces before the pitting process is initiated. However, the oil content of the grease is not affected by the electrical field, but the carboxylate anion stretching and carboxylic group present in the soap residue undergo changes. Figures 6.1 and 6.2 indicate variation of resistivity of lithium base grease 'A' with time at different potential drops across electrodes, and variation as well as 'recouping' of resistivity with time under 50 V

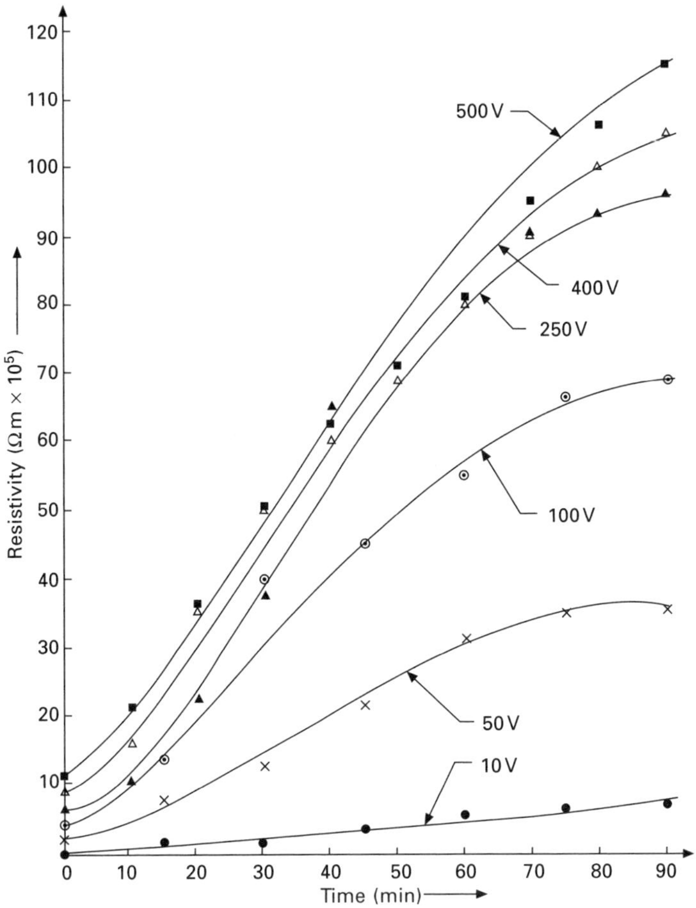

6.1 Variation of resistivity of grease 'A' with time at different potential drops across electrodes.

potential drop across electrodes, respectively. Figures 6.3 and 6.4 show IR spectra of the soap residue of fresh and used greases.[1–5]

The original structure of the grease and structure of the grease after operation of a rolling-element bearing under electrical field have also been studied by X-ray diffraction techniques.[6,7] This indicates that the original structure of the soap residue of the fresh lithium grease is lithium stearate ($C_{18}H_{35}LiO_2$), which changes to lithium palmitate ($C_{16}H_{31}LiO_2$), a lower fraction of hydrocarbon, after operation of a bearing under electrical fields; the original structure of the grease is not changed under pure rolling friction. Also, gamma lithium iron oxide and lithium zinc silicate ($Li_{3.6}Zn_{0.2}SiO_2$) are formed in the presence of Zn, Fe and SiO_2 (Fig. 6.5 and 6.6). In contrast, under pure rolling friction only lithium iron oxide is detected. The crystalline structure

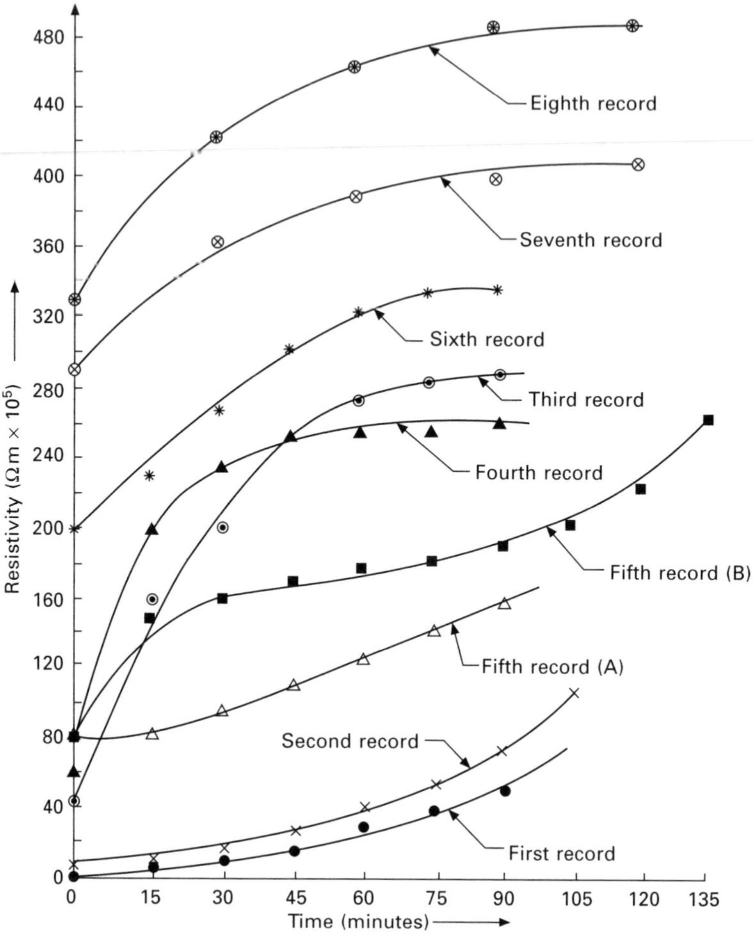

6.2 Variation and recovery of resistivity of grease 'A' with time under 50 V potential drop across electrodes.

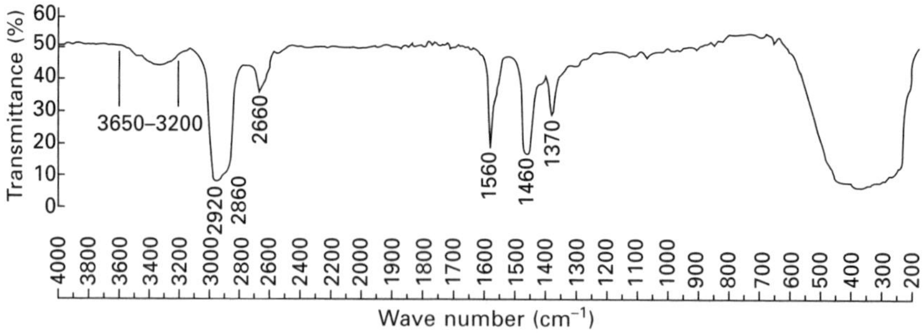

6.3 IR spectrum of soap residue extracted from fresh grease 'A'.

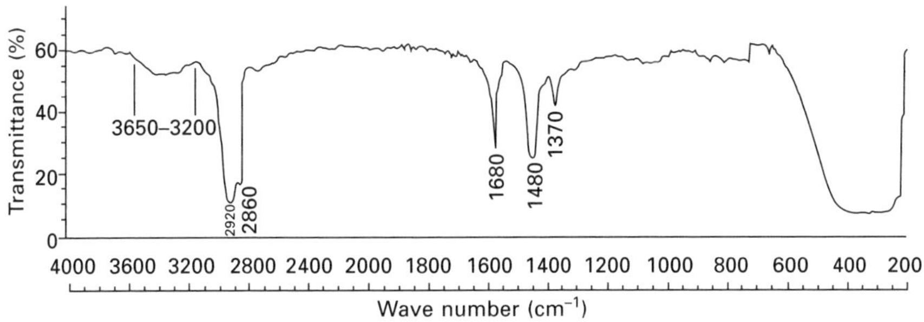

6.4 IR spectrum of grease 'A' taken from NU 326 bearing after exposure to 50 A current (AC) for 41 hours.

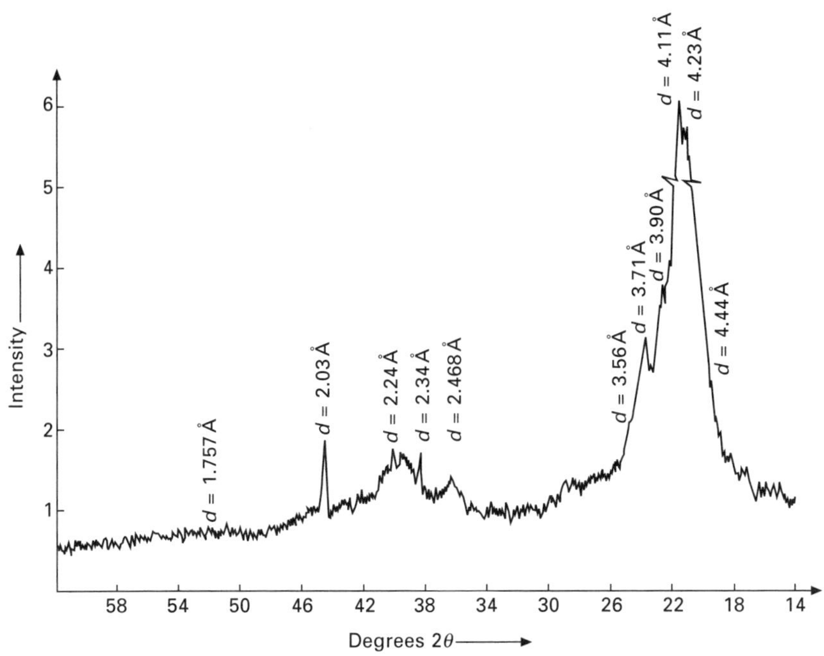

6.5 X-ray diffraction pattern of soap residue for fresh grease 'A'.

of a fresh lithium grease in a bearing changes to an amorphous structure. Under the influence of an electrical current, the formation of lithium hydroxide and lithium carbonate makes the dielectric alkaline and corrodes the bearing surfaces, which lead to increased wear and failure of a bearing.

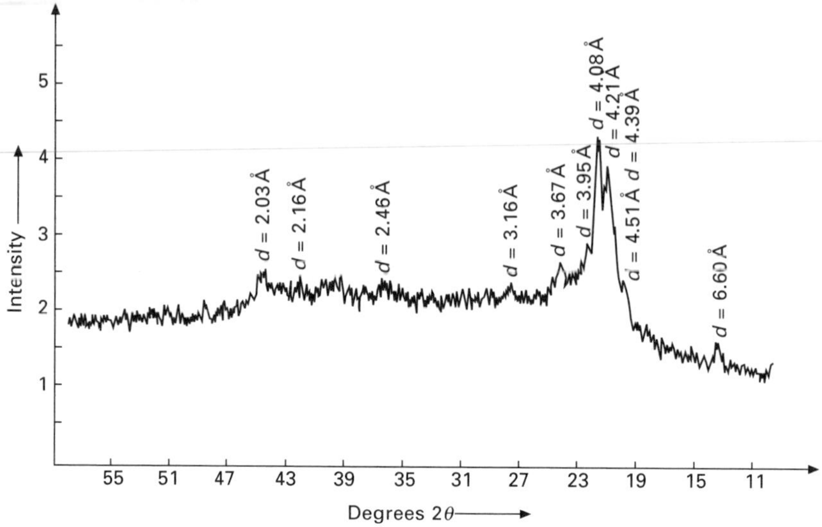

6.6 X-ray diffraction pattern of soap residue of grease from NU 326 bearing after operation under electrical fields for 250 h.

6.4 Effect of current on formation of corrugated patterns on the roller track of races of roller bearings

The passage of current causes local surface heating, which leads to low-temperature tempering, and accelerates formation of corrugations with time (Fig. 6.7). As rolling continues, corrosion increases due to the decomposition of grease 'A' and small particles of material are pulled out from track surfaces at the points of asperity contacts between the races and rollers, which develop scores on the surfaces, besides forming corrugations/flutings on the surfaces. After long operation, the softer tempered surfaces of the races become harder, and thus harder/re-hardened particles due to localized high temperature and load erupt from the craters and intensify the depth of corrugations.

In the presence of low-resistivity lubricant ($10^5 \Omega$ m), in the close asperity contacts between rolling elements and races, current intensity is increased by short-circuiting at the line of contacts that leads to the formation of corrugations in due course. After the formation of corrugations in one half of the races – the resistance becomes higher due to a decrease in contact area – the current leaks through the other half, and thus fully fledged corrugations are formed on the surfaces.

At each revolution of the shaft, part of the circumference of the inner race passes through a zone of maximum radial force, and Hertizian pressure between the rolling elements and raceways (at the line contacts) leads to a maximum shear stress. The maximum shear stress is taken as the criterion

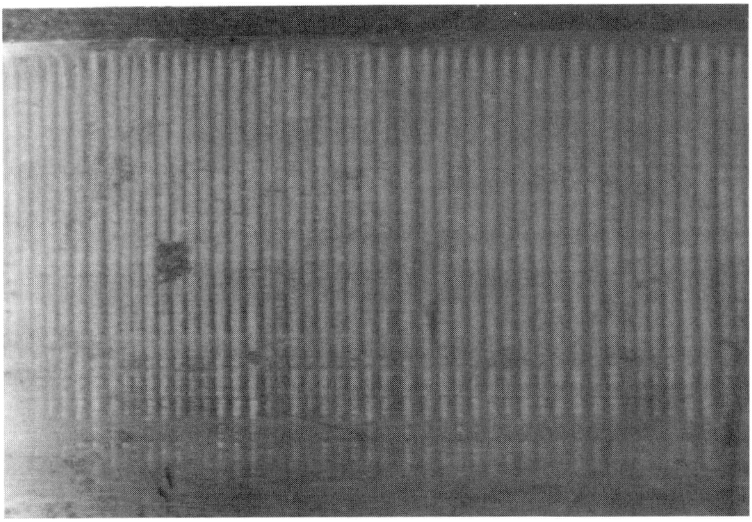

6.7 Corrugation pattern on the inner race of NU 326 motor bearing after about 6000 h of operation.

for yielding and this occurs in the subsurface at a depth approximately equal to half the radius of the contact surface. It is generally at this point that failure of material, if occurring, will initiate. As soon as the fatigue spall appears on the surface, the actual area of the asperity contact between the rolling-element and the race is reduced, which reduces electrical contact resistance and increases flow of current. Furthermore, there is a gradual increase in the width of corrugation by deformation at the asperity contact due to an increase in contact pressure per unit area, which leads to further reduction in electrical contact resistance and increase in flow of high-intensity current. These intensify the corrugation pattern in due course.

When a bearing is significantly loaded, the deformation is made by rolling elements (K) on the races in the loaded zone. The process of deformation, which leads to the formation of a corrugation pattern on the surfaces at the line contacts due to a decrease in contact resistance, is accelerated by the passage of high-intensity current, corrosion and oxidation of surfaces, lubricant characteristics and quality of a bearing.

The pitch of corrugations on the roller tracks depends on the bearing kinematics, the frequency of rotation, the position of plane of action of radial loading, the bearing quality and the lubricant characteristics, and is given as:[8,9]

$$\Delta_{ir} = \pi Dd \,/SF_s Kp(D + d) \qquad [6.1]$$

$$\Delta_{or} = \pi Dd \,/SF_s Kp(D - d) \qquad [6.2]$$

$$\Delta_{re} = \pi d/SF_s\, p \qquad [6.3]$$

The width of corrugations on the surface is not affected by frequency of rotation and depends on the load conditions and bearing kinematics. The width of corrugations on the inner and outer races is given as:[8,9]

$$W_{ir} = 2.15\ [Pd\ (D - d)/pKELD]^{1/2} \quad [6.4]$$

$$W_{or} = 2.15\ [Pd(D + d)/pKELD]^{1/2} \quad [6.5]$$

$$W_{re} = W_{or} + \theta W_{ir} \quad [6.6]$$

where θ, the overlapping coefficient, varies from 0 to 1.

The pitch and width of corrugations are smaller on the inner race than on the outer race. Also pitch and width of corrugations on rollers are affected by corrugation pattern already formed on the races.

6.5 Effect of current leakage on electro-adhesion forces in rolling friction and magnetic flux density distribution on bearing surfaces

The mechanism of adhesion, friction and wear on bearing surfaces in the presence of lubricating film is quite complex. The process of adhesion involves the formation of a junction between the asperities contact, which may finally lead to elastic and plastic deformation under load. Energies of atomic nature are exchanged at the asperities, which may be affected by cage and roller slip due to close interaction of rolling elements with the races.[10]

It is rather difficult to estimate the electro-adhesion forces in the rolling friction. But these can be assessed with a reasonable accuracy by SRV (Schmierstoff –lubricant–material) analysis; the change in coefficient of friction, profile depth and ball scar diameter of the used greases recovered from the active zone of the bearings. This is because of the activity of the zinc additive, i.e. zinc dithiosphosphate or zinc dialkyldithiophosphate (ZDTP) used as a multifunction additive in the grease. Under pure rolling friction it protects the rubbing metal surfaces and contributes to friction and wear reduction, and depends, partly, on the amount of additive absorbed on these surfaces. Physisorption and chemisorption processes precede the chemical reactions with the metal; therefore, it is probable that load-carrying capacity is related to these processes. Correlation between ZDTP adsorption data and wear shows that ZDTP is reversibly physiosorbed on iron at 25 °C, but at 50 °C undergoes chemisorption reactions. On the other hand, decomposition of ZDTP in the lithium base greases under the influence of electrical fields leads to the formation of lithium zinc silicate ($Li_{3.6}Zn_{0.2}SiO_2$) in the presence of high relative percentage of free lithium and silica impurity in the grease under a high temperature in the asperity contacts along with the formation of gamma lithium iron oxide. Besides this, the original structure of lithium stearate changes to lithium palmitate. These changes are not detected under pure rolling friction.

The above changes in the lubricating medium of the bearings operated under electrical fields are reflected in SRV analysis, and are related to electro-adhesion forces in the rolling friction. Also, these are contributed by the medium–metal interaction, rate of chemical reaction, affinity of lubricating medium components for the metal, availability of free metal and temperature rise. If the grease recovered from such bearings is put in a new bearing, the bearing may fail prematurely owing to higher temperature rise even under pure rolling friction.[11]

It is established that under the influence of an electrical field, a roller bearing using low-resistivity grease develops alternating magnetic flux density distribution on the inner race and rolling elements (Fig. 6.8).[11] The maximum magnetic flux density on the inner race has been detected up to 95 G, and on rolling elements up to 18 G. However, significant flux density is not developed on the surfaces of ball bearings. Also, a bearing using high-resistivity grease does not develop significant flux density distribution on its surfaces.[12–15] Besides this, there occur changes in electro-adhesion forces when bearings are lubricated with low-resistivity grease ($10^5\,\Omega$ m) rather than grease of high resistivity ($10^9\,\Omega$ m). Pure rolling friction does not greatly affect the electro-

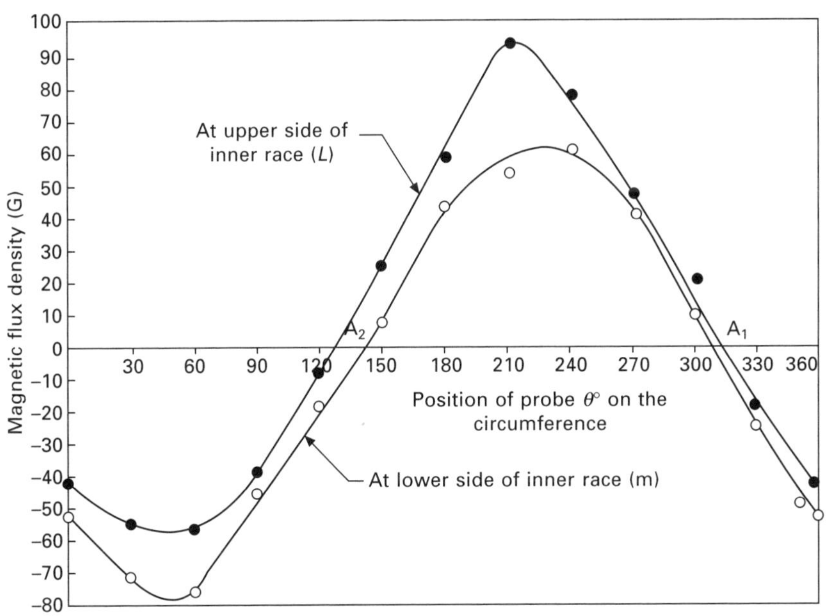

6.8 Magnetic flux density distribution around the track surface of inner race of NU 326 bearing after passing 50 A (AC) at 1.2 to 2.3 V for 250 h (bearing using grease 'A').

adhesion forces. In general, by the study of magnetic flux density distribution along with the study of damaged and corrugated bearing surfaces, and also by the analysis of deterioration of greases used in bearings,[1,5] a diagnosis of leakage current through a non-insulated bearing of an electrical motor can be established.[11–15]

6.6 Effect of operating parameters on the threshold voltages and impedance response of non-insulated rolling-element bearings

Investigations reveal that the first and second threshold voltages (V_{t1} and V_{t2}) appear under the influence of electrical currents in the bearings, depending on the lubricant resistivity, oil film thickness, bearing conditions and the operating parameters. The detected threshold voltages are primarily responsible for momentary flow of current and the further increase in current intensity with a slight change in potential drop across the bearings (Fig. 6.9). The impedance of the bearings becomes negligible as the current intensity across the bearing increases (Fig. 6.10). However, the impedance is more affected by the speed and film thickness than by the load on the bearings.

The threshold voltages (V_{t1} and V_{t2}) decrease as the load on the bearing is gradually increased at constant speed. It is found that the threshold coefficients (V_{t1k} and V_{t2k}) are almost constant at a particular operating speed and the change in load does not affect the threshold coefficients. The threshold coefficients are given by:[16]

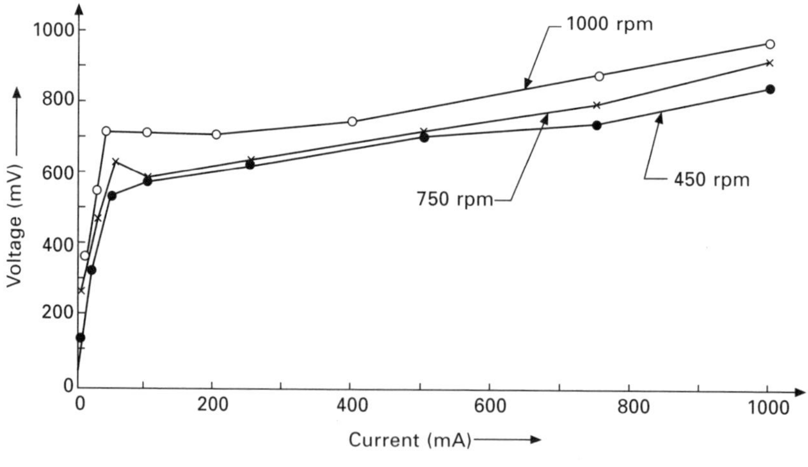

6.9 Variation of bearing current with voltage of NU 330 bearing using lubricant 'B' at different speeds at 750 kgf load.

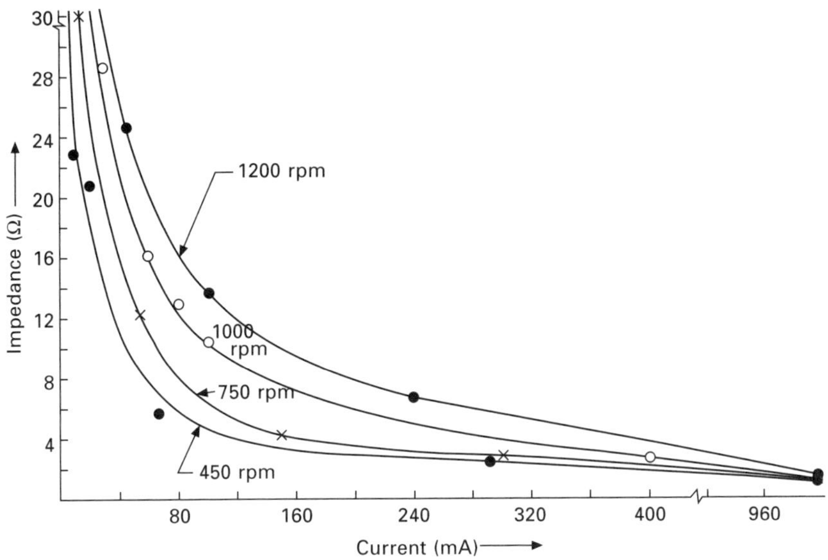

6.10 Variation of bearing impedance with current at different speeds and 450 kgf load – for NU 330 bearing (using lubricant 'B').

$$V_{t1k} = V_{t1}P^{0.3} \qquad [6.7]$$

$$V_{t2k} = V_{t2}\,P^{0.3} \qquad [6.8]$$

In general, the impedance of bearings varies with the parameters of operation and resistivity of the lubricant. The variation in load at constant speed has very little effect on the bearing impedance at the fixed levels of current intensity compared with the change in speed at fixed load. Also, an increase in current intensity reduces the bearing impedance very significantly irrespective of the operating parameters. For reliable operation, the safe limit of the potential drop across the bearing elements should be less than the first threshold voltage.

6.7 Impedance, capacitance and charge accumulation on roller bearings

Under the influence of potential drop across a roller bearing, the minimum film thickness between the races and the rollers offers maximum capacitance and minimum capacitive reactance depending on the permittivity of the lubricant. The electrical interaction between the races and the rollers in the presence of the oil film is like a resistance capacitor (RC) circuit and offers an impedance to the current flow.[17]

In short, under different operating conditions, the capacitance between the outer race and a roller is higher than that between the inner race and a roller. The capacitance and resistance of a bearing depend on the film thickness and width of deformation, and are governed by the permittivity and resistivity of the lubricant. The capacitance and resistance of races can be determined by the following formulae:[17]

$$C_{ir} = 2\xi L(\beta h_0)^{-1/2} \tan^{-1} [0.5\ W_{ir}(\beta / h_0)^{1/2}] \quad [6.9]$$

$$C_{or} = 2\xi L(\delta h_0)^{-1/2} \tan^{-1}[0.5 W_{or}\ (\delta / h_0)^{1/2}] \quad [6.10]$$

$$R_{ir} = \frac{\rho(\beta h_0)^{1/2}}{2L \tan^{-1}[0.5\, W_{ir}\, (\beta / h_0)^{1/2}]} \quad [6.11]$$

$$R_{or} = \frac{\rho(\delta h_0)^{1/2}}{2L \tan^{-1}[0.5\, W_{or}\, (\delta / h_0)^{1/2}]} \quad [6.12]$$

The equivalent bearing capacitance is given as:

$$C_b = K/W(X_{cir} + X_{cor}) \quad [6.13]$$

where

$$X_{cir} = \frac{1}{WC_{ir}} \text{ and } X_{cor} = \frac{1}{WC_{or}} \quad [6.14]$$

Stored charges in the bearing depend on bearing capacitance and charge increases with applied voltage provided the dielectric is not dissociated. The stored charge in a bearing is given as:

$$Q = VC_b \quad [6.15]$$

It is established that the equivalent capacitance of a bearing decreases with increasing speed at constant load but increases with load at constant speed (Fig. 6.11). Also, a bearing lubricated with high-resistivity lubricant as opposed to low-resistivity lubricant, with the same permittivity, behaves like a capacitor up to the first threshold voltage. In addition to this, for a bearing to accumulate charges, the ratio of capacitive reactance to active resistance should be less than unity.

6.8 Contact temperature, contact stresses and slip bands initiation on roller track of races

Current passing through a bearing at the line contacts between the roller tracks and rollers, and the corresponding impedance generates heat and increases the temperature instantaneously. This increases the contact stresses and enables the determination of the number of cycles before the slip bands are initiated on the track surfaces of a bearing lubricated with low-resistivity lubricant ($10^7\,\Omega$cm).[18]

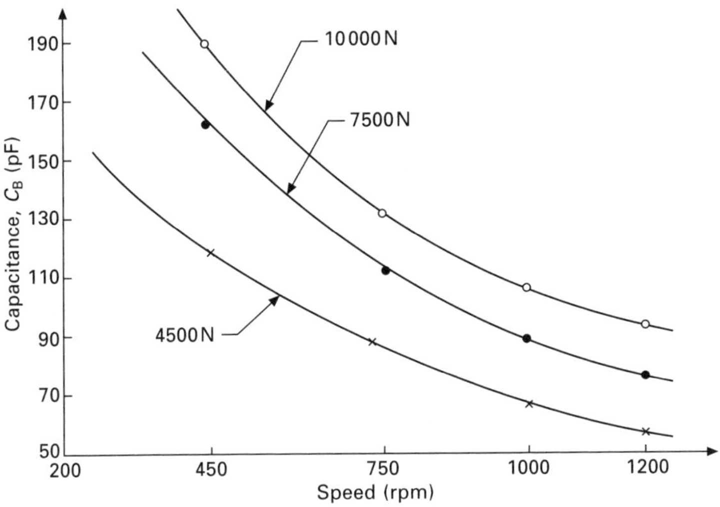

6.11 Variation of capacitance with speed at different radial loads in NU 330 type bearing using a high-resistivity lubricant 'B'.

The duration of contact between roller tracks of races and a roller has been theoretically determined. This depends on the pitch diameter and the roller diameter of a bearing. It is higher for the roller track of the outer race and a roller than for the roller track of the inner race and a roller. It decreases with rotating speed frequency but increases with width of contact. The values can be determined as:

$$t_{ir} = 2DW_{ir}/\pi F_s(3D + d)(D - d) \quad [6.16]$$

and

$$t_{or} = 2DW_{or}/\pi F_s(D^2 - d^2) \quad [6.17]$$

The temperature rise at each line contact between the roller track of races and a roller in each shaft rotation depends on the bearing kinematics, the depth of slip bands, the number for rollers in the loaded zone and the properties of the bearing material. It decreases with an increase in frequency of rotation but increases with the resistance of the bearing and with the current. The temperature rise of the inner and outer race tracks is given as:[18]

$$T_{irn} = T_{orn} = I^2R_bK/\pi F_s dLH\varepsilon C \quad [6.18]$$

Stresses on the roller tracks are determined as:

$$\alpha_{irn} = -\frac{\alpha E(T_{irn} - T_a)}{1 - \mu} \quad [6.19]$$

It is established that the slip bands (Fig. 6.12) on the roller track of races of the NU 326 bearing at a current of 50 A are initiated after fewer than 41 h of

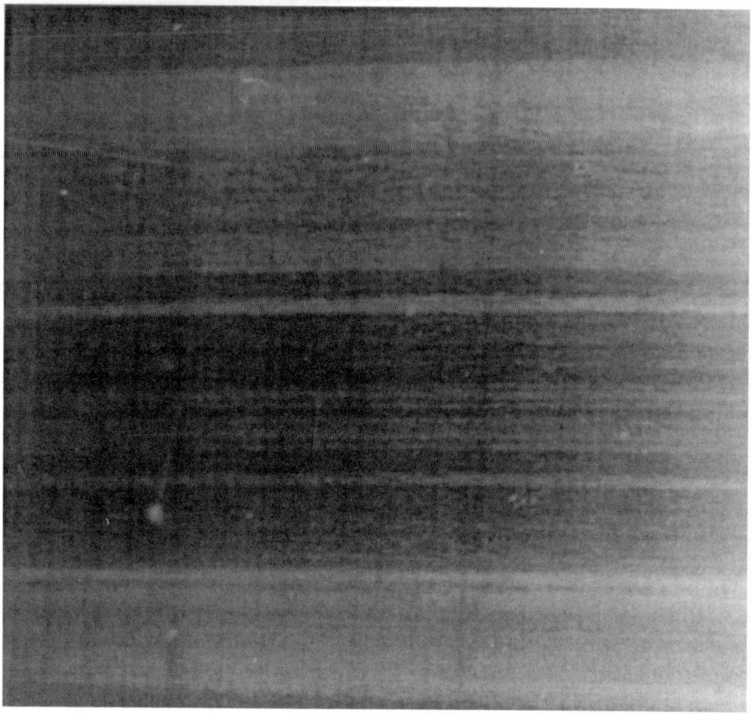

6.12 Enlarged view of slip bands on inner race of NU 326 roller bearing after exposure to a current of 50 A (AC) for 41 h (lubricant A).

operation, 15.15 h or 10^6 cycles after the bearing temperature is stabilized. At a lower current intensity, more time elapses before slip bands appear on the roller tracks.

6.9 Effects of instantaneous charge leakage on roller tracks of roller bearings lubricated with high-resistivity lubricants

The effects of instantaneous charge leakage on the rise of contact temperature between rollers and roller track of races of roller bearings lubricated with high-resistivity lubricant have been analysed. Estimates of the leakage of charge between roller tracks and rollers during momentary asperity contacts along with an expression for the instantaneous contact resistance between roller and race have been used to establish heat generated and instantaneous temperature rise in the contact zone on the roller tracks in each shaft rotation. Using this temperature rise, the contact stresses are determined and the

6.13 Damaged inner race of NU 2215 bearing (using lubricant 'B') under electrical fields.

minimum number of cycles calculated before craters appear on the roller track of the races (Fig. 6.13).

The instantaneous temperature rise due to sudden charge leakage during each contact of a roller with the roller track of inner race and outer race is determined as:[19]

$$T_{ir} = \frac{4Q^2 D}{\pi F_s C_b^2 L_a R_s H \varepsilon C(3D + d)(D - d)} \quad [6.20]$$

$$T_{or} = \frac{4Q^2 D}{\pi F_s C_b^2 L_a R_s H \varepsilon C(D^2 - d^2)} \quad [6.21]$$

A bearing with a higher capacitance and less charge accumulation takes a longer time before the craters are initiated on the roller tracks.

6.10 Capacitive effects of roller bearings on repeated starts and stops of a machine

Recent investigations give the time required for the charge accumulation and increase of charge with time on the bearing surfaces based on the bearing capacitance, the resistance of film thickness and the shaft voltage. Also, the investigation gives the effect of gradual leakage of the accumulated charges with time as the shaft voltage falls as soon as the power supply to the machine is put off.

The ratio of contact cycles required for the charge accumulation and gradual discharge of the accumulated charges on the bearing surfaces depending on the bearing to shaft voltage has been analysed. The number of cycles and number of repeated starts and stops before initiation of craters on roller track of races, as against the bearing to shaft voltage, have been theoretically established to restrict the deterioration and damage of the bearings.

The time required in developing charges (Q_{ir} and Q_{or}) on a roller and roller track of inner and outer races is given as:[20]

$$T_{cir} = -\, C_{ir}\, R_{ir} \log_e (1 - a) \qquad [6.22]$$

$$T_{cor} = -\, C_{or}\, R_{or} \log_e (1 - a) \qquad [6.23]$$

Similarly, the time required to discharge the accumulated charges from roller track of inner as well as outer races and a roller/rollers is given as:[20]

$$T_{dir} = -\, C_{ir}\, R_{ir} \log_e a \qquad [6.24]$$

$$T_{dor} = -\, C_{or}\, R_{or} \log_e a \qquad [6.25]$$

where $a = V/V_o$ (ratio of bearing to shaft voltages).

The number of starts and stops before initiation of craters on the roller track of inner and outer races (N_{ssi} and N_{sso}) can be determined as:[20]

$$N_{ssi} = -\frac{C_{si}}{F_s C_{ir} R_{ir} \log_e a(1 - a) 20 F_s} \qquad [6.26]$$

and

$$N_{sso} = -\frac{C_{so}}{F_s C_{or} R_{or} \log_e a(1 - a) 20 F_s} \qquad [6.27]$$

It is shown that with an increase of bearing to shaft voltage, the number of starts and stops to initiate craters on the roller track of races decreases. For the NU 330 bearing, the number of starts and stops decreases from 803.63 to 463.50 as the ratio of bearing to shaft voltage increases from 0.5 to 0.9 (Fig. 6.14 and 6.15).

6.11 Mechanism of bearing failures

When current leaks through the roller bearing in which low-resistivity lubricant has been used, a silent discharge passes through the bearing elements. This creates magnetic flux density distribution on the surfaces. Initially, this is accompanied by electrochemical decomposition of the grease and corrosion on the bearing surfaces and then gradual formation of slip bands at the line contacts, which lead to the formation of flutings and corrugations, and subsequently wear increases and the bearing fails.[18] In contrast, when high-resistivity lubricant is used, the charges accumulate on the bearing surfaces due to polarization till they reach the threshold critical value at which the feeble current is conducted through a bearing,[19] which is not able to result in significant flux density distribution on the surfaces. However, mass transfer at the elevated local temperature accompanies this in a few cycles of operation

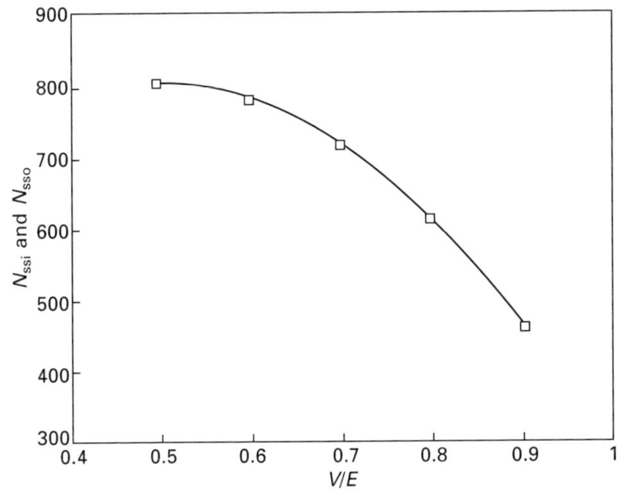

6.14 Variation of number of starts and stops of a motor before formation of craters on roller track of inner and outer races (N_{ssi} and N_{sso}) at various levels of bearing the shaft voltage *V/E* of the NU 330 bearing operating under the influence of electrical currents.

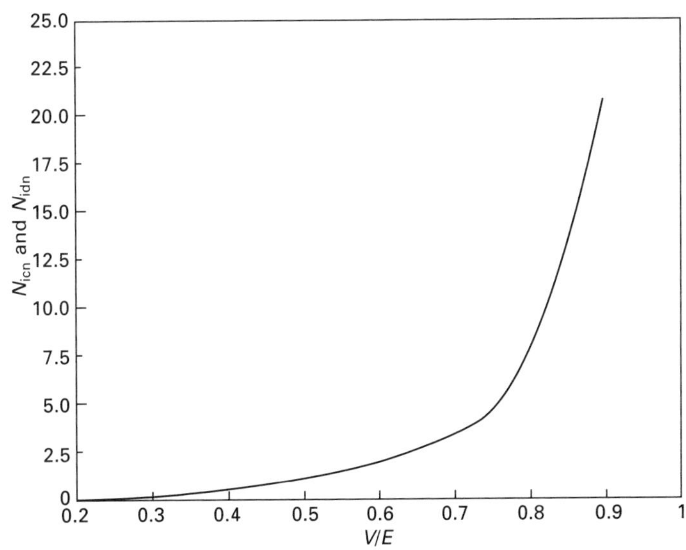

6.15 Variation of the ratio of shaft revolutions to accumulate and discharge of accumulated charges (N_{icn}/N_{idn}) at various levels of bearing to shaft voltages (*V/E*) on roller tracks of inner and outer races of a roller bearing operating under the influence of electrical current.

on asperities of the interacting surfaces by instantaneous discharge. This reduces the fatigue life and initiates crater formation on the surface by the welding effect and causes failure of a bearing.

6.12 Conclusion

Recent investigations on responses and performance of bearings under the influence of shaft voltage have given insight into bearing behaviour, deterioration of lubricant and lubricant decomposition. Besides this, failure of bearing can be established under bearing current. The effect of bearing capacitance on charge accumulation and of electro-adhesion forces on the magnetic flux density distribution is understood. Also, contact temperature and number of cycles before the slip bands and craters initiation on roller tracks, leading to the reduction in the life of bearing operating under the influence of shaft voltages, have been theoretically established. Analysis of the safe limit of a number of starts and stops of a machine influencing the bearing deterioration and crater formation acts as a potential tool to diagnose the bearing performance under different levels of bearing to shaft voltages. But the problem remains as to how to increase bearing life under the effect of shaft voltage. This may be achieved by the improved lubricant characteristics, bearing design or by alteration of bearing material properties. These aspects require further investigation to complete the 'know-how' gap. However, bearing insulation is the only solution presently available to tackle this problem.

6.13 References

1 Prashad, H. and Murthy, T.S.R., 'Behaviour of Greases in Statically Bounded Conditions and When Used in Non-insulated Anti-friction Bearings Under the Influence of Electrical Fields', *Lubr. Eng.*, **44**(3), 239–246, 1988.

2 Prashad, H., 'Experimental Study on Influence of Electrical Fields on Behaviour of Grease in Statically Bounded Conditions and When Used in Non-insulated Bearings', *BHEL J.*, **7**(3), 18–34, 1996.

3 Polacios, J.M., 'Elasto-hydrodynamic Films of Lithium Greases', *Macanique, Materiaux, Electricite (GAMI)*, 176, 365–366, 1980 (also published in *NLGI Spokesman*, March 1981).

4 Pastnikov, S.N., '*Electrophysical and Electrochemical Phenomena in Friction, Cutting and Lubrication*', VNR, 1978.

5 Prashad, H., 'Variation and Recovery of Resistivity of Greases – An Experimental Investigation', *J. Lubr. Sci. (France)*, 11–1, 73–103, November 1998.

6 Prashad, H., 'Diagnosis of Lithium Greases Used in Rolling-element Bearings by X-Ray Diffractometry', *STLE Trans*, **32**(2), 205–214, 1989.

7 Prashad, H. and Murthy, T.S.R., 'Deterioration of Lithium Greases Under the Influence of Electrical Current – An Investigation', *J. Lubr. Sci. (France)*, 10–4, 323–342, August 1998.

8 Prashad, H., 'Investigations of Damaged Rolling-element Bearings and Deterioration

of Lubricants Under the Influence of Electrical Fields', *Wear*, 176, 151–161, 1994.

9 Prashad, H., 'Investigations of Corrugated Pattern on the Surfaces of Roller Bearings Operated Under the Influence of Electrical Fields', *ASME/ASLE Tribology Conference*, San Antonio Marriott (5–8, Oct. 1987), also published in *Lubr. Eng.*, **44**, 710–718. 1988.

10 Prashad, H., 'The Effect of Cage and Roller Slip on the Measured Defect Frequency Response of Rolling-element Bearings', *ASLE Trans*, **30**(3), 360–367, July 1987.

11 Prashad, H., 'The Effects of Current Leakage on Electroadhesion Forces in Rolling Friction and Magnetic Flux Density Distribution on the Surface of Rolling-element Bearings', *Trans. ASME, J. Tribol.*, **110**, 448–455, 1988.

12 Tasker, J.L. and Graham, R.S., 'Effects of Magnetic Flux on Rolling-element/Bearings', *IEE Conference*, 17–19, Sept. 1985, Publication 254, pp. 152–156.

13 Prashad, H., 'Magnetic Flux Density Distribution on the Track Surface of Rolling-element Bearings – An Experimental and Theoretical Investigation', *Tribol. Trans*, **39**(2), 386–391, 1996.

14 Prashad, H., 'Determination of Magnetic Flux Density on the Surfaces of Rolling-element Bearings as an Indication of the Current That Has Passed Through Them – An Investigation', *Tribol. Int.*, **32**, 455–467, 1999.

15 Prashad, H., 'Determination of Magnetic Flux Density on the Surfaces of Rolling-element Bearings – An Investigation', *BHEL J.*, **21**(2), 49–66, August 2000.

16 Prashad, H., 'Effects of Operating Parameters on the Threshold Voltages and Impedance Response on Non-insulated Rolling-Element Bearings under the Influence of Electrical Currents', *Wear*, **117**, 223–240, 1987.

17 Prashad, H., 'Theoretical Analysis of Impedance, Capacitance and Charge Accumulation of Roller Bearings Operated Under Electrical Fields', *Wear*, **125**, 223–239, 1988. Also in *BHEL J.*, **9**(2), 21–30, 1988.

18 Prashad, H., 'Analysis of the Effects of Electrical Currents on Contact Temperature, Residual Stresses, and Slip Bands Initiation on Roller Tracks of Roller Bearings', *Wear*, **131**, 1–14, 1989.

19 Prashad, H., 'Theoretical Analysis of the Effects of Instantaneous Charge Leakage on Roller Bearings Lubricated with High Resistivity Lubricants under the Influence of Electric Current', *Trans. ASME, J. Tribol.*, **112**, 37–43, 1990.

20 Prashad, H., 'Theoretical Analysis of Capacitive Effect of Roller Bearings on Repeated Starts and Stops of a Machine Operating under the Influence of Shaft Voltages', *Trans. ASME, J. Tribol.*, **114**, 818–822, October 1992.

6.14 Nomenclature

a	ratio of potential difference across bearing to shaft voltage
C	specific heat of bearing material
C_b	equivalent bearing capacitance
C_{ir}, C_{or}	capacitance between inner race and a roller, and outer race and a roller, respectively
C_{si}, C_{so}	number of cycles before initiation of craters on roller track of inner race, and outer race, respectively

d	diameter of rolling element
D	pitch diameter
E	Young's modulus of elasticity
F_s	shaft rotational frequency
h_o	minimum oil film thickness
H	depth of crater/slip bands on roller track of races
I	bearing current
K	number of rolling elements in the loaded zone
L	length of rolling element
L_a	summation of the length of asperity contact on circumference of roller track during contact with a roller
N_{icn}	number of shaft revolutions to accumulate charge
N_{idn}	number of shaft revolutions to discharge accumulate charge
N_{ssi}, N_{sso}	number of starts and stops before the formation of craters on roller track of inner race and outer race, respectively
p	position of plane of action of radial loading ($p = 1, 2, 3, ...$)
P	resultant load on bearing
Q	stored electrical charge
Q_{ir}	charge on roller track of inner race
Q_{or}	charge on roller track of outer race
r	radius of rolling element
R	outer radius of inner race
R_i	inside radius of outer race
R_b, R_s	equivalent bearing resistance under operating and static conditions, respectively
R_{ir}, R_{or}	resistance between roller track of inner race and a roller, and outer race and a roller, respectively
S	bearing coefficient(s)
t_{ir}, t_{or}	duration of each line contact between roller track of inner race and a roller, and outer race and a roller, respectively
T_a	ambient temperature
T_{dir}, T_{dor}	time required to discharge the accumulated charges from roller track of inner race and a roller, and outer race and a roller, respectively
T_{ir}, T_{or}	instantaneous temperature rise of roller track of inner race and outer race, respectively, due to charge leakage during each contact with a roller
T_{irn}, T_{orn}	instant temperature rise of roller track of inner race and outer race, respectively, in each shaft rotation
V	applied voltage/voltage across bearing
V_o	shaft voltage
V_{t1}, V_{t2}	first and second threshold voltages, respectively
V_{t1k}, V_{t2k}	first and second threshold voltage coefficients, respectively

W_{ir}, W_{or}	width of corrugations on roller track of inner race and outer race, respectively
W_{re}	width of corrugations on rolling elements
W	$2\pi f$
X_{cir}, X_{cor}	capacitive reactance between inner race and a roller, and outer race and a roller, respectively
ρ	resistivity of lubricant
α	coefficient of thermal expansion
α_{irn}, α_{orn}	tangential stress on roller track of inner race and outer race, respectively
μ	Poisson ratio
ε	density of bearing material
ξ	permittivity (dielectric constant) of lubricant
Δ_{ir}, Δ_{or}	pitch on corrugations of roller track of inner race and outer race, respectively
Δ_{re}	pitch of corrugations of the surface of rolling elements
θ	overlapping coefficient
β	0.5 $[1/r + 1/R]$ inner race constant
δ	0.5 $[1/r - 1/R_i]$ outer race constant

7

Effect of oil grades and clearance ratios on the reliability of cylindrical hydrodynamic bearings

7.1 A general review

In this study, the effect of oil grades and clearance ratios on reliable performance, safe operation and design of steady-loaded, pressure-fed, hydrodynamic cylindrical bearings is highlighted. A procedure for the assessment of bearing performance has been developed based on the maximum oil temperature in the load-carrying oil film, variation of oil viscosity with temperature and effective oil temperature in the bearing.

Viscosity coefficients are obtained, using an iterative procedure, from the viscosity–temperature relationship for different grades of oil. The viscosity integral is evaluated as a function of inlet and outlet oil temperatures in the load-carrying oil film, using splines. Nomographs are plotted to estimate oil viscosity–temperature in the load-carrying oil film as a function of viscosity integral, and also to determine the transition speed as a function of oil viscosity, clearance ratio and shaft diameter. The viscosity integral is directly evaluated by the bearing parameters for various operating conditions. Safe load-carrying capacity and maximum oil temperature in the bearing using different grades of oil under similar operating conditions and clearance ratios are compared. The theoretical load-carrying capacity at different eccentricity ratios, thus obtained, is compared with the published data and found to match for eccentricity ratios between 0.6 and 0.75.

7.2 Introduction

Reliable operation of a bearing necessitates a supply of an adequate amount of lubricant to the surfaces under relative motion. The quantity of oil required depends on the amount of generated heat transferred within the bearing. There is a considerable temperature difference between the inlet and outlet of the load-carrying oil film. It is well established that the total flow of lubricant from the supply oil grooves is divided into two streams in the bearing.[1] One stream is drawn into the clearance by the rotation of the journal, while the other stream is forced into the bearing by the delivery supply pressure. The lubricant forced into the bearing helps to expel the hot circulating oil in the vicinity of the groove. The oil drawn into the clearance existing across the film thickness cools the bearing and determines the effective temperature and viscosity of the oil in the bearing. However, if the total flow from the supply grooves is greater than the flow required across the film thickness, the excess flow does not contribute to any additional cooling of the bearing, but only cools the outlet oil flow. The bearing will be starved if the total flow from the supply grooves is less than that required across the film thickness.[2] Design procedures based on the mean of the load-carrying oil film inlet and outlet temperatures are not sufficiently accurate, since the rise of temperature within this film, where there is intensive wear of the bearing, is not taken into account. A temperature as high as 120° may be developed in this region. Hence, the temperature in the loaded zone can be considered as a diagnostic tool for optimum reliable operation of the bearing.[3]

Under similar operating conditions, the temperature rise of the oil in the loaded zone when the speed is increased without changing load is comparatively less than that obtained by increasing load on the bearing without changing the shaft speed. This is because of a decrease in relative eccentricity and a corresponding increase in the film thickness, which causes comparatively more heat transfer by the increased flow of oil from the loaded to the unloaded zone, and gives rise to a higher oil temperature in the unloaded zone of the bearing when the shaft speed is increased. In contrast, with an increase in load on the bearing, the relative eccentricity increases, which increases the temperature rise in the loaded zone, and under such conditions less hot oil from the loaded zone enters the unloaded zone because of the decrease in film thickness. This causes a lower temperature rise in the unloaded zone because the fresh oil from the supply groove is mixed in, but increases in homogeneity in the oil temperature, since the total inlet flow of oil to the bearing in both cases is constant. That is why, for loaded bearings, it is essential to determine the oil flow considering the oil temperature at the outlet of the load-carrying part of the oil film.

Considering the above mechanism of oil flow in the bearings, this study gives an easy adaptable design procedure and diagnostic technique for performance evaluation of the cylindrical bearings under various parametric and operating conditions.

7.3 Background

The governing equation for temperature rise in the load-carrying oil film has been worked out by Nika[4] for a finite bearing. The oil viscosity varies with film thickness and the operating parameters of the bearing. This equation is very complex and coefficients in the equation can be found only by the method of approximation. The approximate temperature of the oil film is worked out by integrating the energy equation on the assumption that the amount of heat generated by friction is carried out by oil flow and only part of it is transmitted through the bearing housing.

The governing equation for computation of oil temperature in the loaded zone is given as:[4]

$$\rho C_v U \frac{\partial T}{\partial x} = A\mu\left(\frac{\partial V_x}{\partial y}\right)^2 + \lambda\left(\frac{\partial^2 T}{\partial x^2} + \frac{\partial^2 T}{\partial y^2}\right) \qquad [7.1]$$

The second term of Equation (7.1) on the right hand side represents the heat transfer through the bearing housing. Ignoring this term (as explained above), the simplified governing equation for a stabilized regime in a plane parallel to the stream of incompressible oil may be represented as:

$$\rho C_v U \frac{\partial T}{\partial x} = A\mu\left(\frac{\partial V_x}{\partial y}\right)^2 \qquad [7.2]$$

On integrating, the following equation is obtained:[3]

$$\int_1^2 \frac{\partial T}{\mu} = \left(\frac{UL}{427\rho C_v}\right)\left(\frac{4}{d^2\psi^2}\right)\left[\frac{(2+\varepsilon^2)}{\varepsilon(1-\varepsilon^2)}\right] \times \left[4\log\xi + \frac{6}{\xi} - \frac{1.5}{\xi^2}\right]_{\xi_1}^{\xi_2} \qquad [7.3]$$

where

$$\xi_1 = \frac{2+\varepsilon^2}{2(1+\varepsilon)}$$

$$\xi_2 = \frac{2+\varepsilon^2}{2(1-\varepsilon)}$$

and

$$U = \pi dn/60$$

Equation (7.3) determines the change of oil temperature along the load-carrying part of oil film. Limits of integration of $\int \partial T/\mu$ indicate the temperature of the oil at the inlet and the outlet of the load-carrying oil film. The function $\int_1^2 \partial T/\mu$, which is the viscosity integral, is worked out separately in the following sections.

7.4 Theoretical

7.4.1 Viscosity–temperature relationship

To determine oil temperature in the load-carrying part of the oil film, the following viscosity–temperature relationship is used:[3]

$$\mu = \exp(C + D/T)^2 \qquad [7.4]$$

This relationship is more accurate than the other relations normally used, since the viscosity coefficients C and D are independent of temperature in the oil film. Differentiating Equation (7.4), gives:

$$\partial\mu/\mu = -2\left(C + \frac{D}{T}\right)\frac{D}{T^2}\,\mathrm{d}T \qquad [7.5]$$

Temperature in the oil film is given by:

$$T = \frac{D}{(\ln\mu)^{0.5} - C} \qquad [7.6]$$

From Equation (7.6), it is evident that:

$$\int\frac{\partial T}{\mu} = -0.5\int\left(\frac{D}{(\ln\mu)^{0.5} - C]^2[\mu^2(\ln\mu)^{0.5}}\right)\partial\mu \qquad [7.7]$$

or

$$\int\frac{\partial T}{\mu} = -\int\left[\frac{\mathrm{f}(\mu)}{\mu^2}\right]\partial\mu \qquad [7.8]$$

where

$$\mathrm{f}(\mu) = 0.5\left\{\frac{D}{[(\ln\mu)^{0.5} - C]^2[(\ln\mu)^{0.5}]}\right\} \qquad [7.9]$$

7.4.2 Statistical method for determination of viscosity coefficients

Equation (7.4) can be written as:

$$(\ln\mu)^{0.5} = C + \frac{D}{T} \qquad [7.10]$$

If $(C + D/T_{\mathrm{i}})$ is represented by M_i, then:

$$\sum_{i=1}^{n}\left(M_{\mathrm{i}} - C + \frac{D}{T_{\mathrm{i}}}\right) = 0 = S$$

On differentiating, it is evident that:

$$\partial S/\partial C = \Sigma\left(M_i - C + \frac{D}{T_i} \right) = 0$$

and

$$\frac{\partial S}{\partial D} = \Sigma\left(\frac{1}{T_i} \right)\left(M_i - C + \frac{D}{T_i} \right) = 0$$

Hence:

$$C = \overline{M} - D(\overline{1/T}) \qquad [7.11]$$

$$D = \frac{\overline{M}(\overline{1/T}) - (\overline{M/T})}{(\overline{1/T})^2 - (\overline{1/T^2})} \qquad [7.12]$$

where $\overline{M}$, $(\overline{1/T})$ and $(\overline{1/T^2})$ are the average values of the respective parameters.

7.5 Evaluation of viscosity coefficients

The values of viscosities at various temperatures are given in Fig. 7.1 for oil grades 'A_1', 'B_1' and 'C_1'. Values of viscosity coefficients 'C' and 'D' are determined independently for these grades of oil using Equation (7.4), by an iterative procedure, which is an extension of the Gauss Newton method for solving the coefficients in a multivariable non-linear equation. The coefficients are evaluated in the temperature range of 50 to 125 °C. The statistical method is also followed, using Equations (7.11) and (7.12), to determine the values of viscosity coefficients and compare these with those evaluated by Gauss Newton method. The values of coefficients are found to be closely comparable. The values determined by the Gauss Newton method are given in Table 7.1.

A nomograph (shown in Fig. 7.1) is drawn using the viscosity–temperature variation of oil grades 'A_1', 'B_1' and 'C_1', and the respective values of viscosity coefficient 'C', so as to enable evaluation of coefficient 'C' for any unknown grade of oil. To determine the value of coefficient 'C' for an unknown grade of similar oil, firstly a curve showing variation of oil viscosity with temperature is drawn, similar to that shown in Fig. 7.1. Then an identical procedure, as shown in Fig. 7.1, is followed to determine the coefficient 'C'. However, the value of coefficient 'D' can be taken as 1180, if the lubricant is of the same origin and has the same base oil and additives as the oils 'A_1', 'B_1' and 'C_1'.

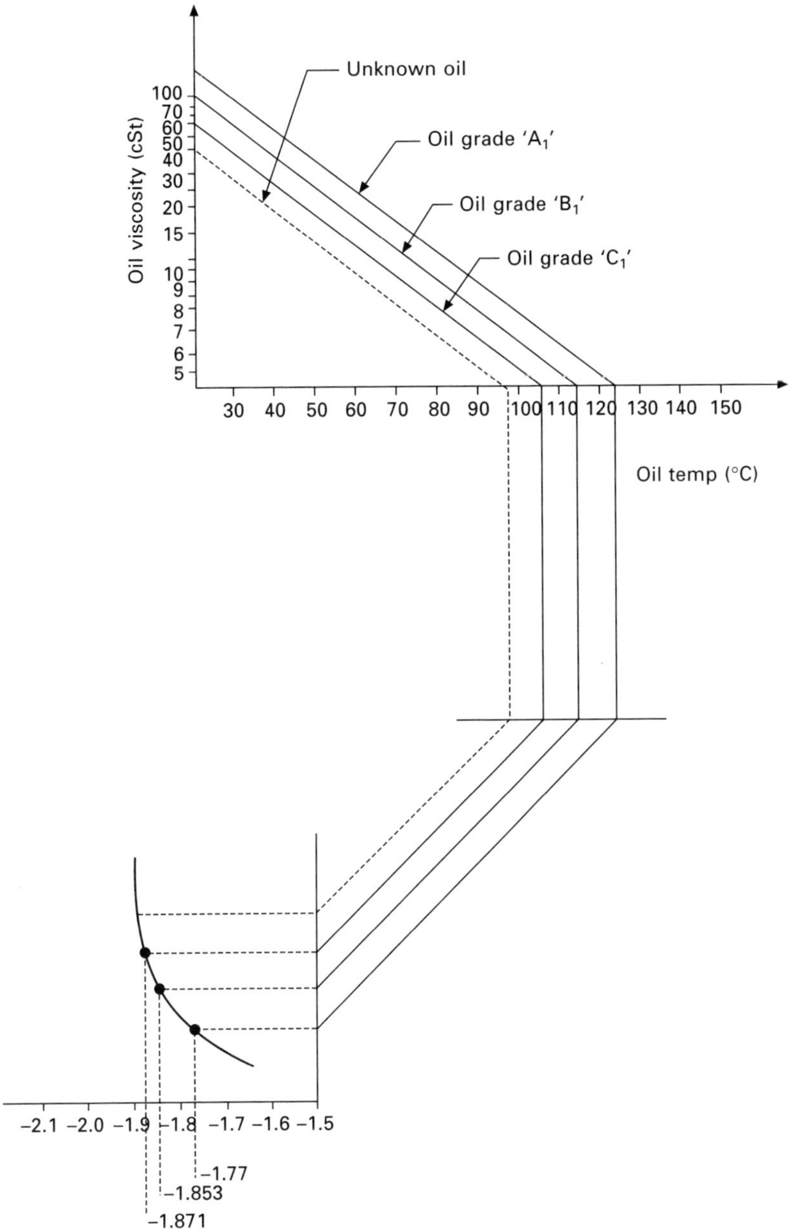

7.1 Estimation of value of viscosity coefficient 'C' for different grades of oil (1cSt = 10^{-6} m^2 s^{-1}).

Table 7.1 Values of viscosity coefficients

Oil grade	Coefficient 'C'	Coefficient 'D'
A_1	−1.770	1180
B_1	−1.853	1180
C_1	−1.871	1180

7.6 Determination of viscosity integral

7.6.1 Theoretical approach

The values of the viscosity integral $\int \partial T/\mu$, i.e. $-\int[\mathrm{f}(\mu)/\mu^2)\partial\mu]$ has been computed for 'A_1', 'B_1' and 'C_1' grades of oils, using Equation (7.7). The methodology of splines was adopted to evaluate the integral of a function. For each grade of oil, the lower limit of viscosity integral was taken as 60 °C assuming that the oil enters the load-carrying oil film at this temperature. To evaluate the viscosity integral, values of coefficients 'C' and 'D' (as shown in Table 7.1) were used for the respective grades of oil. The values of viscosity integral with reference to the viscosity of the oil from the load-carrying oil film, for oil grades 'A_1', 'B_1' and 'C_1', are presented in Table 7.2.

7.6.2 Functional nomographs

The nomographs in Fig. 7.2, 7.3 and 7.4 were plotted to facilitate evaluation of the viscosity integral with reference to viscosity of the oil in the load-

Table 7.2 Variation of viscosity integrals with viscosity of oils

Serial No.	Oil grade					
	A_1		B_1		C_1	
	Oil viscosity μ (cP)	Viscosity integral	Oil viscosity μ (cP)	Viscosity integral	Oil viscosity μ (cP)	Viscosity integral
1	3.465	11.119	3.509	10.322	3.292	10.885
2	4.245	7.946	3.894	8.570	3.369	8.998
3	4.938	6.198	4.332	7.104	3.985	7.653
4	5.457	5.247	4.938	5.620	4.331	6.485
5	6.064	4.371	5.414	4.735	4.981	4.934
6	6.930	3.486	6.064	3.828	5.371	4.187
7	7.797	2.817	6.714	3.078	6.064	3.236
8	9.096	2.095	7.667	2.298	6.714	2.440
9	9.962	1.764	8.663	1.707	7.797	1.606
10	11.695	1.231	10.396	1.03	8.663	1.070
11	13.860	0.803	11.695	0.634		
12	15.593	0.540				

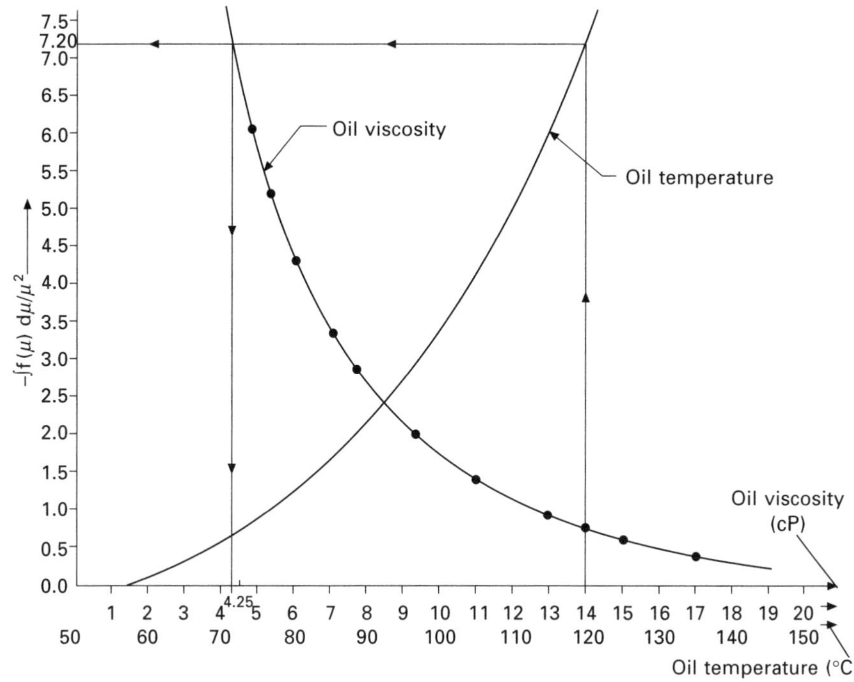

7.2 Variation of viscosity integral with oil temperature and viscosity for oil grade 'A'.

carrying part of the oil film for oil grades 'A_1', 'B_1' and 'C_1', respectively. The respective oil temperature corresponding to viscosity of the oils, determined from Equation (7.6), is also plotted in the functional nomographs.

Considering Babbitt metal temperature and oil characteristics as well as the oil oxidation stability, the maximum permissible oil temperature for a short duration in the load-carrying oil film is 120 °C, for reliable and safe operation of the bearing,[5] whereas the maximum allowable limit of bulk oil temperature is 70 °C.[6]

The horizontal line corresponding to 120 °C (shown in each nomograph) indicates the limiting acceptable viscosity integrals in the load-carrying oil film. Above this value, the bearing operational parameters or oil grade has to be changed for reliable operation. The critical values of the viscosity integrals at 120 °C are 7.20, 9.70 and 11.20 for 'A_1', 'B_1' and 'C_1' grades of oil (Fig. 7.2, 7.3 and 7.4) respectively. Viscosity integrals, as indicated in Equation (7.3), depend upon various bearing parameters including eccentricity and clearance ratios, speed of operation, length of load-carrying oil film. The variation of bearing parameters affects the viscosity integrals, which in turn, influences the bearing performance.

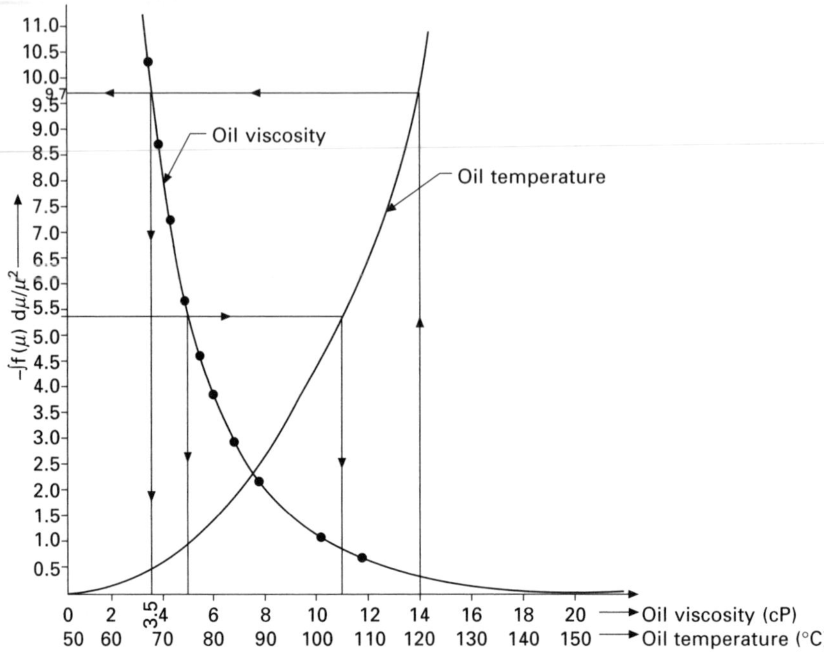

7.3 Variation of viscosity integral with oil temperature and viscosity for oil grade B_1.

7.7 Assessment of bearing performance

To assess the bearing performance, the value of eccentricity ratio is determined using the following empirical relation:[3]

$$\bar{S} = \frac{2.08}{(1 - \varepsilon)} \tag{7.13}$$

where $\bar{S}$, the loading coefficient, is given by:[3]

$$\bar{S} = \frac{2p\psi^2}{\mu_e w} \quad \text{and} \quad p = \frac{\bar{S}\mu_e w}{2\psi^2} \tag{7.14}$$

In Equation (7.14), $w = 2\pi n/60$ and $p = P/ld$.

The effective viscosity (μ_e) is calculated through effective temperature in the bearing by the following relation:[2]

$$T_e = \frac{T_{max} - T_1}{2} + T_1 \tag{7.15}$$

First, the viscosity integrals are determined independently using Equation (7.3) for given parameters and operating conditions of the bearing. Then T_{max} is estimated against the viscosity integrals, for oil grades 'A_1', 'B_1' and

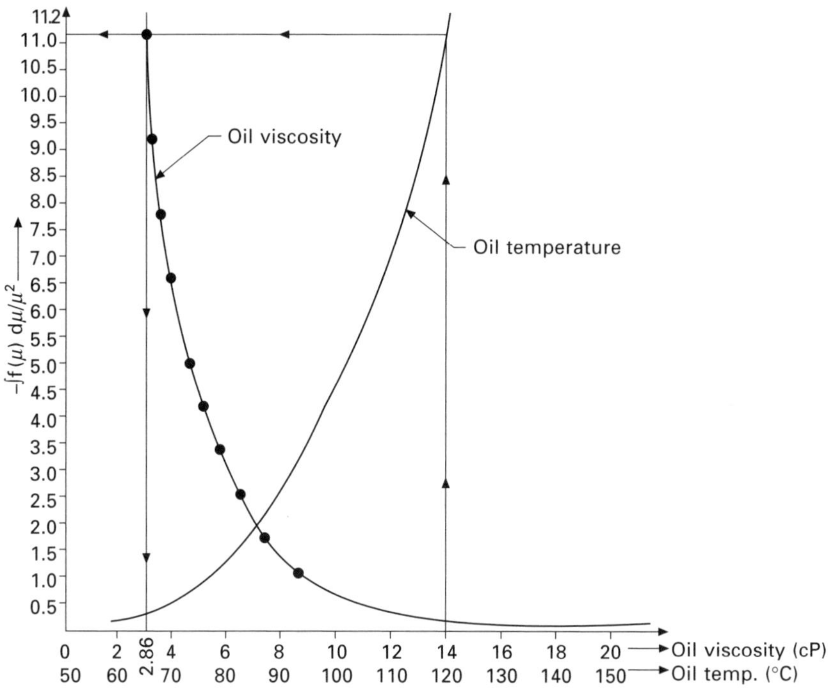

7.4 Variation of viscosity integral with oil temperature and viscosity for oil grade C_1.

'C_1', using nomographs given in Fig. 7.2 to 7.4. It may be noted that 'L', i.e. the peripheral length of load-carrying oil film, is very difficult to estimate. A theoretical solution for the determination of the boundary limit for the length of the load-carrying oil film has not yet been correctly established. Experimental work carried out by Petrunichev[7] revealed that with an increase in the load coefficient, the angle pertaining to the load-carrying oil film reduces. To assess bearing performance, an average length of load-carrying oil film (L) corresponding to angle 90° ($L = \pi\, d / 4$) may be reasonably used for eccentricity ratios ranging from 0.6 to 0.8.[7]

On the basis of maximum temperature in the oil film, the relative comparison of the bearing performance under various operating conditions can be established. Data showing variation of clearance and eccentricity ratios with viscosity integrals have been computed for different values of speed (rpm); the variation at 2500 rpm is shown in Fig. 7.5. The maximum values of viscosity integral for different oil grades are also indicated in Fig. 7.5, for easy determination of optimum values of clearance and eccentricity ratios at a speed of 2500 rpm.

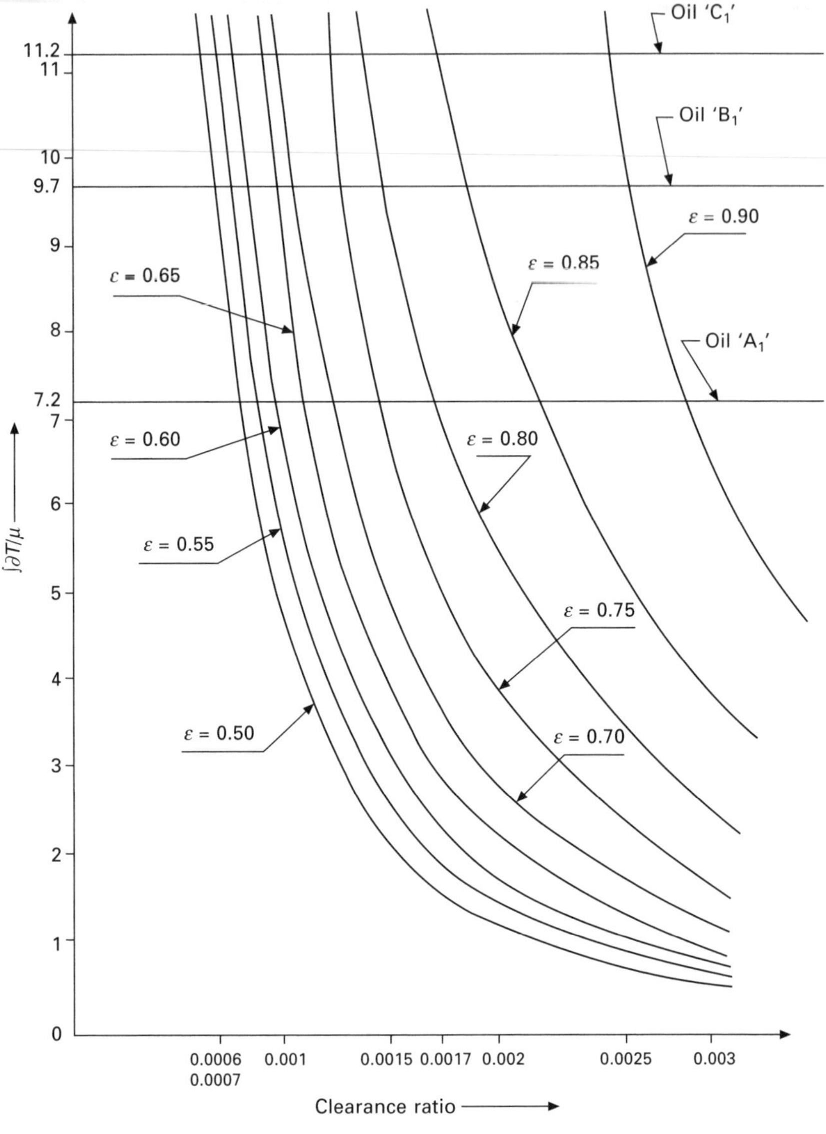

7.5 Variation of viscosity integral with clearance and eccentricity ratios of bearings with 2500 rpm.

7.8 Effect of oil grades on temperature rise and safe load-carrying capacity of bearings

Using Equation (7.3) and Fig. 7.2 to 7.4, the temperature-rise in the load-carrying oil film under various operating speeds (1500 to 4000 rpm) was calculated for different grades of oil for the bearing having $C_d/d = 0.0015$

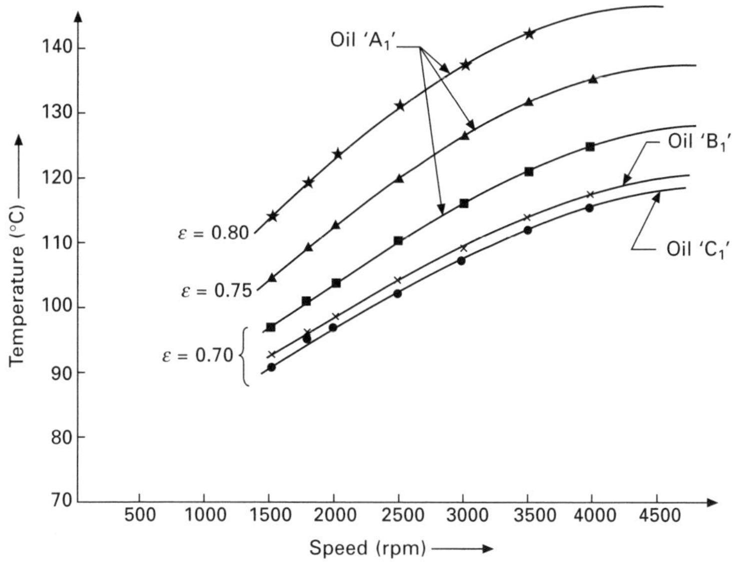

7.6 Variation of temperature of oil at outlet of load-carrying oil film at varying operating speeds and eccentricity ratios of bearing with clearance ratio of 0.0015.

and $\varepsilon = 0.7$ (the oil enters the load-carrying oil film at 60 °C). Effects of oil grades are also analysed under different eccentricity ratios (0.5 to 0.8) for the bearing operating at 2500 rpm having $C_d/d = 0.0015$. The temperature rise in the load-carrying oil film under these conditions is shown in Fig. 7.6 and 7.7, for oil grades 'A_1', 'B_1' and 'C_1'.

The optimum safe load-carrying capacity (at 120 °C oil temperature in the loaded zone) at different operating speeds and clearance ratios against viscosity integrals of 7.20, 9.70 and 11.20 was worked out using Fig. 7.2 to 7.4, Equations (7.13) and (7.14), and is shown in Fig. 7.8 to 7.10 for oil grades 'A_1', 'B_1' and 'C_1', respectively.

7.9 Bearing turbulence and transition speed

The operating characteristics of a cylindrical bearing at high surface speeds have revealed abnormalities beyond a certain critical value. A rapid increase in bearing torque, power loss and oil film temperature occurred as speed was increased beyond the critical value, while oil flow decreased below normal. These phenomena were attributed to the onset of instability of turbulence in the bearing oil film with an accompanying increase in energy absorption within the film.[8] Under the turbulent conditions, the frictional coefficient was considerably higher than that predicted by the friction versus load parameter curve, leading towards the higher oil film temperature of the bearing. The

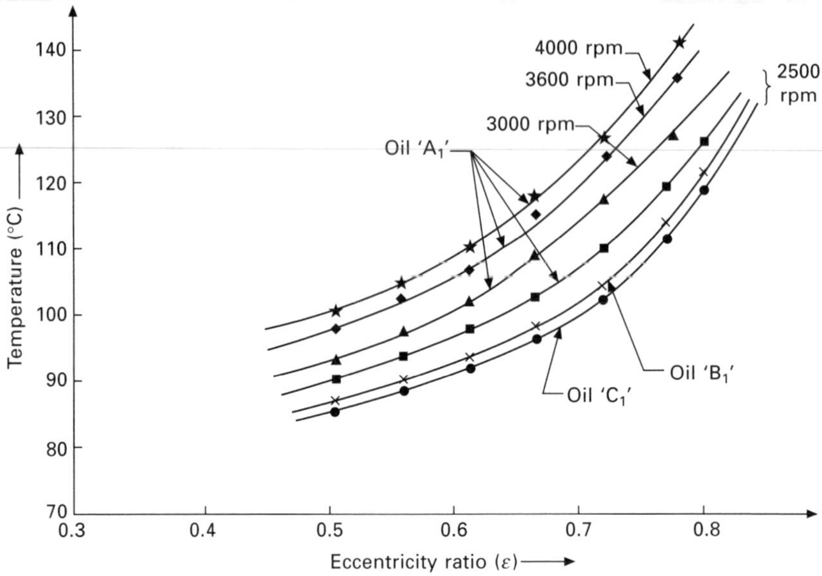

7.7 Variation of temperature of oil at outlet of load-carrying oil film at varying eccentricity ratios and at different operating speeds of bearing with clearance ratio of 0.0015.

turbulence in the bearing occurred when the Reynolds number increased beyond the critical value. The Taylor criterion for the flow between rotating cylinders was applied to determine the transition speed in the cylindrical bearings, at which the laminar regime changed to turbulent regime.

The Taylor criterion for the critical point beyond which instability occurs may be written as:[9]

$$R_{crt} = 41.1\ (d/C_d)^{0.5} \qquad [7.16]$$

For a turbulent regime, $R_e > R_{crt}$ and at transition speed, $R_e = R_{crt}$. It follows that:

$$R_e = \frac{\pi N d C_d}{2/v} = 41.1\left(\frac{d}{C_d}\right)^{0.5} \qquad [7.17]$$

Hence,

$$N = \frac{0.0016 v}{\psi^{1/2} d^2} \qquad [7.18]$$

where N (transition speed) is in rpm, v in centistokes (1cSt = 10^{-6}m^2/s) and d in metres.

From Equation (7.18), it is evident that for a bearing to operate in the laminar regime, a change of clearance ratio, oil viscosity or bearing diameter

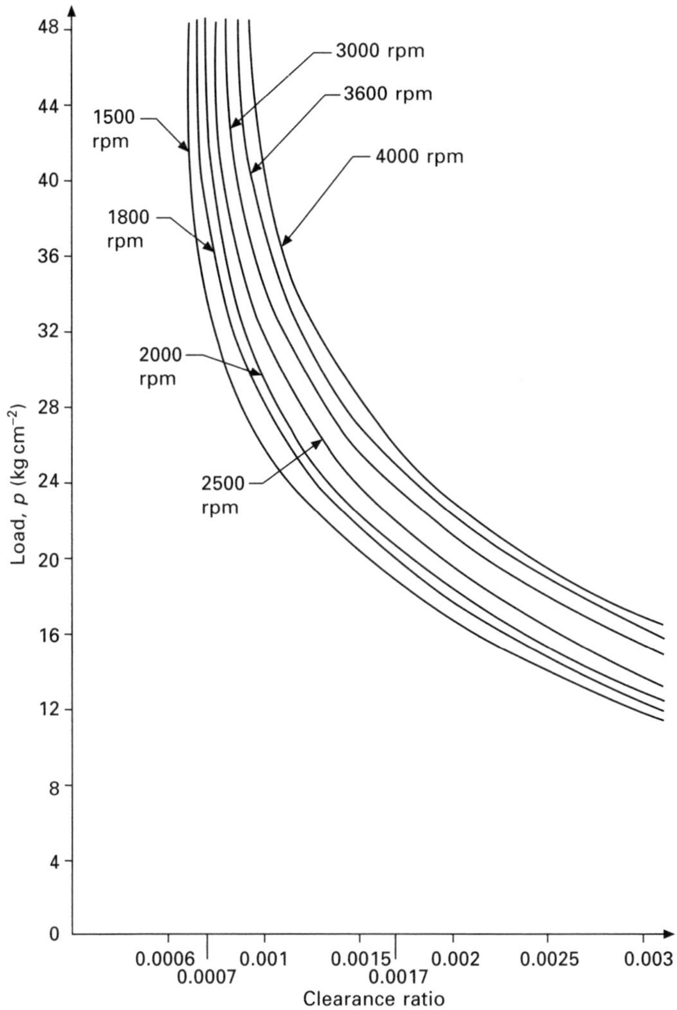

7.8 Variation of safe load-carrying capacity with clearance ratio at various speeds of operation using oil grade A_1.

or the introduction of a floating sleeve design must be considered to keep the power loss and temperature rise within the permissible limits.

For the immediate assessment of the turbulent regime, nomographs indicating transition speed with reference to clearance ratio, shaft diameter and oil viscosity, shown in Fig. 7.11 can be readily used, before applying the design procedure, which is valid only for the laminar regime.

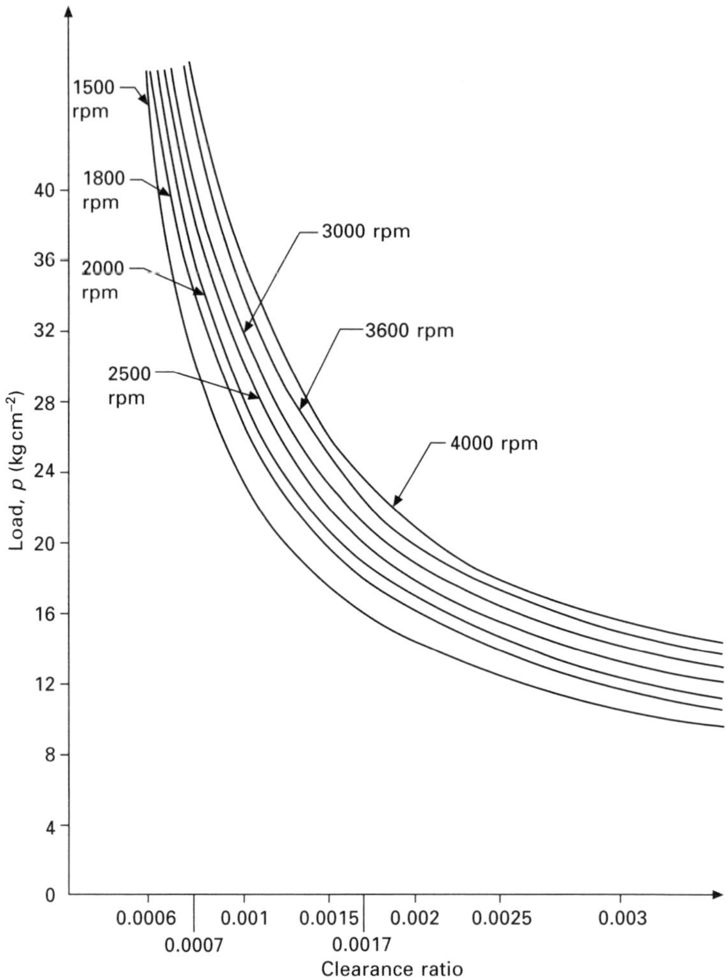

7.9 Variation of safe load-carrying capacity with clearance ratio at various speeds of operation using oil grade B_1.

7.10 Results and discussion

7.10.1 Inter-relationship between viscosity integral, transition speed and clearance ratio

The viscosity coefficient '*C*' was significantly affected for oils more viscous than oil grade 'B_1' as compared with oils less than oil grade 'B_1' (Fig. 7.1). This was because of the rate of change in viscosity with temperature. The viscosity coefficient *D* remained constant (i.e. 1180) for different grades of oil.

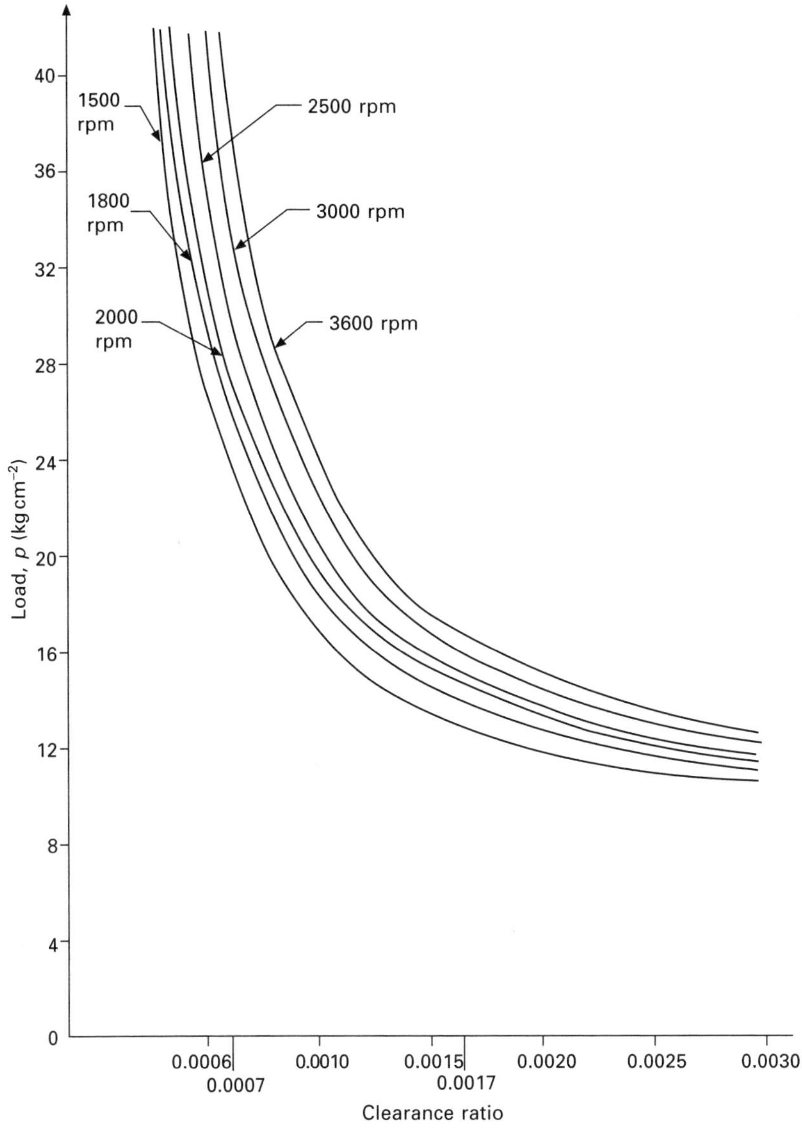

7.10 Variation of safe load-carrying capacity with clearance ratio at various speeds of operation using oil grade C_1.

It is evident from Fig. 7.2 to 7.4 that the viscosity integral increased as the viscosity of the oil decreases, and the rate of variation became asymptotically high for a viscosity generally below 3 cP, whereas it was negligible for viscosity above 17 cP. Even a slight reduction in oil viscosity below 4 cP affected the viscosity integral for oil grade 'A_1' more significantly than in the case of oil grades 'B_1' and 'C_1'. For 'A_1', as the oil viscosity reduced from 4 to 3.5 cP

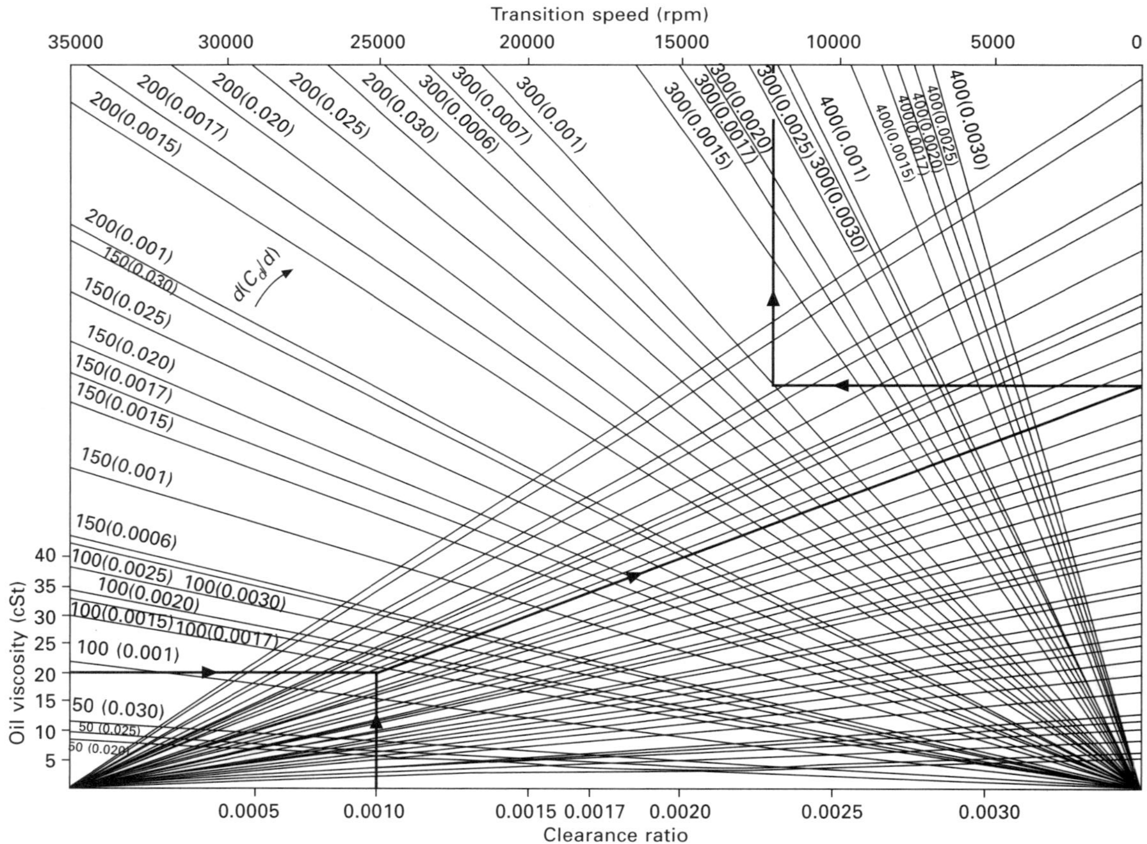

(a)

7.11 Nomograph indicating transition speed with clearance ratio, shaft diameter and oil viscosity (a) 50–400 mm shaft diameter; (b) 250–800 mm shaft diameter.

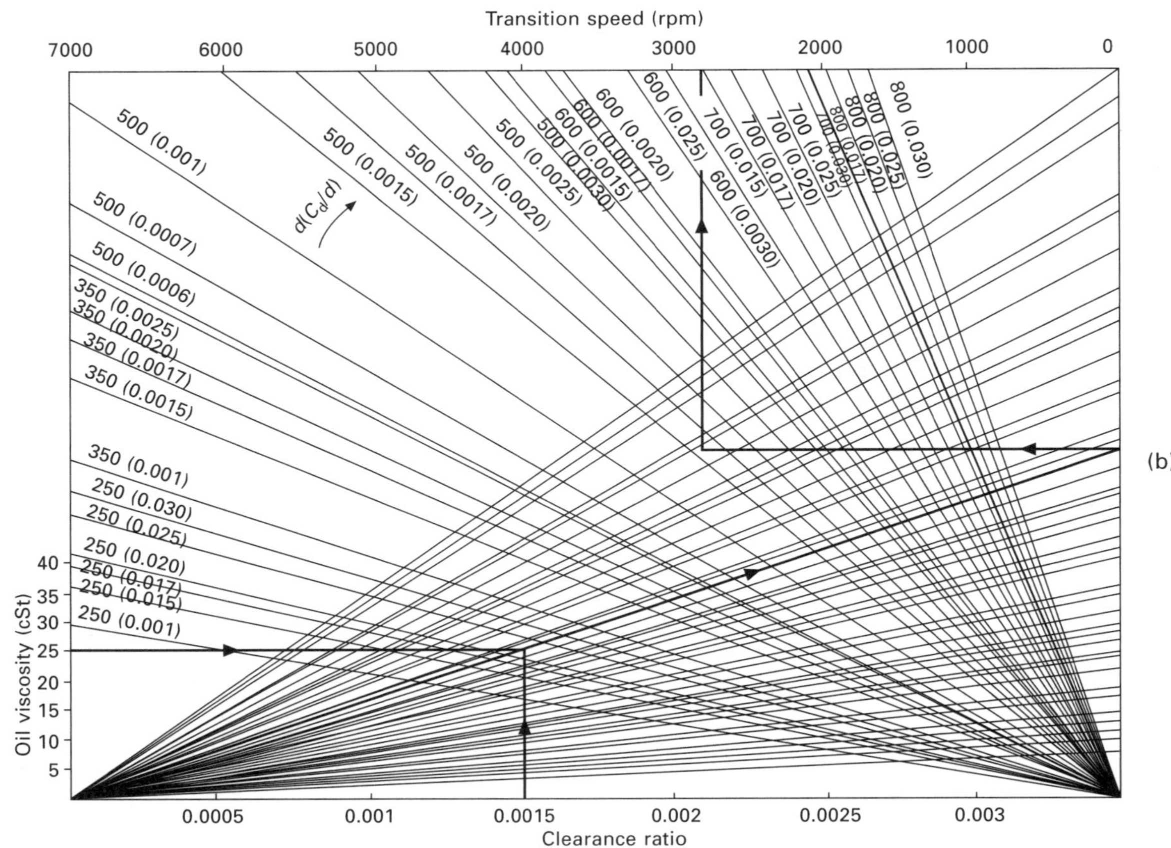

7.11 (Continued)

in the load-carrying oil film, the viscosity integral increased from 7.50 to 10 (beyond the curve range) as shown in Fig. 7.2, while for 'B_1' and 'C_1', the viscosity integral increased only from 9.2 to 9.7 and 6.2 to 7.4, respectively (Fig. 7.3 and 7.4). The change in optimum clearance ratio at a given speed of operation and at the desired eccentricity ratio was relatively negligible for viscosity integrals above 9.7 (Fig. 7.5).

The transition speed decreased with a decrease in viscosity (Fig. 7.11) for a given bearing clearance ratio, whereas the viscosity integral increased with a decrease in viscosity. The increase in viscosity integral indicates that the bearing was operating with less viscous oil and higher clearance ratio (Fig. 7.2 to 7.5). These changes decreased the load-carrying capacity and initiated turbulence in a bearing at a lower speed of operation. The transition speed of 11 250 rpm for a bearing of 300 mm diameter operating with 0.001 clearance ratio and 20 cSt oil viscosity changed to 5600 rpm when the oil viscosity was reduced to 10 cSt at the same clearance ratio, as shown in Fig. 7.11(a). From Fig. 7.5, it is evident that at a speed of 2500 rpm, a bearing designed for an eccentricity ratio of 0.7 with a clearance ratio of 0.0017 had a viscosity integral of 3.9, whereas at the same eccentricity ratio, if the bearing was to operate with a viscosity integral of 5, the optimum requirement of clearance ratio would be 0.0015.

From Fig. 7.8 and 7.9, it is evident that at a clearance ratio of 0.0017, if the existing oil grade 'A_1' became less viscous and its viscosity became equivalent to that of oil grade 'B_1' then its load-carrying capacity reduced from 22 kg cm^{-2} (Fig. 7.8) to 21 kg cm^{-2} (Fig. 7.9) at a speed of 2500 rpm. To obtain the same optimum load-carrying capacity, the clearance ratio must be changed from 0.0017 to 0.0015 or a different grade of oil should be used.

For the same rise of temperature in the load-carrying oil film, the viscosity integral was higher for oil grade 'C_1' than for 'B_1' and 'A_1'. For the maximum allowable oil temperature, the values of viscosity integrals were 11.2, 9.7 and 7.2 for oil grades 'C_1', 'B_1' and 'A_1', respectively, and the corresponding viscosities of oils were 2.86, 3.5 and 4.25 cP (Fig. 7.2, 7.3 and 7.4). This pattern of viscosity integrals shows that at a given eccentricity ratio, the optimum value of clearance ratio is lower for oil grade 'C_1' than for oil grades 'B_1' and 'A_1', for identical operating conditions. At a speed of 2500 rpm, the optimum value of clearance ratio for maximum safe load-carrying capacity for bearing operating with an eccentricity ratio of 0.7 was 0.00128 for oil grade 'A_1' for a viscosity integral of 7.2. Similarly, clearance ratios were 0.00114 and 0.00112 for viscosity integrals of 9.7 and 11.2, respectively, for oil grades 'B_1' and 'C_1', respectively (Fig. 10.5). The greater the difference in viscosity integrals in different grades of oil, the higher will be the difference in the optimum clearance ratios of the bearings under identical conditions of operation. This also explains the variation of the optimum bearing clearance ratio with the viscosity of oils.

7.10.2 Effect of oil grades on temperature rise

From Fig. 7.6, it is evident that the temperature in a load-carrying oil film in the bearing using oil grade 'A_1' increased more significantly than when oil grades 'B_1' and 'C_1' were used, with an increase in speed, for the same clearance and eccentricity ratios. For a clearance ratio of 0.0015 and an eccentricity ratio of 0.7, the rise of temperature in load-carrying oil film was 27.5, 24.5 and 22 °C for oil grades 'A_1', 'B_1' and 'C_1', respectively, on varying the operating speed from 1500 to 4000 rpm. On operating the bearing with a clearance to diameter ratio of 0.0015 at 2500 rpm, with an increase in the eccentricity ratio from 0.5 to 0.8, the variation of temperature in oil film was found to be 35 and 34.5 °C for oil grades 'A_1' and 'B_1', respectively, as against 32 °C for oil grade 'C_1' (Fig. 7.7). This shows that the increase in eccentricity ratio under the same speed of operation leads to a higher rise in temperature in the oil film than is found with an increase in speed under the same eccentricity and clearance ratios for bearings, using different grades of oils. From this, it can be concluded that the oil film thickness in the bearing was more affected by the operating speed than the load on the bearing, which led to increased oil flow in the load-carrying oil film and a reduction in temperature rise at higher speed. The maximum difference between the rise in temperature under any eccentricity ratio for $C_d/d = 0.0015$ at speed of 2500 rpm was about 4.5 °C using oil grades 'A_1' and 'B_1' as against 1.5 °C using oil grades 'B_1' and 'C_1', respectively (Fig. 7.7). The same pattern was valid for other operating speeds and this may be correlated with the behaviour of viscosity integrals for different grades of oils.

7.10.3 Effect of oil grades on safe load-carrying capacity

From Fig. 7.8 to 7.10, it is evident that with an increase in clearance ratio, the safe load-carrying capacity per unit area of the bearing was reduced. With an increase in clearance ratio, the transition speed of the bearing also reduced (Fig. 7.11). Under identical conditions, the bearing having the same clearance ratio has a greater load-carrying capacity (p) at higher speed than at lower speed of operation. This is attributed to an increase in oil film thickness with increase in operating speed. Load-carrying capacity under the same operating conditions increased from 16 to 23 $kg\,cm^{-2}$, 14.8 to 21 $kg\,cm^{-2}$ and 10.80 to 16.7 $kg\,cm^{-2}$, for the bearing operating in the range of 1500–4000 rpm with a clearance ratio of 0.002 and using oil grades 'A_1' , 'B_1' and 'C_1', respectively.

At identical speed and C_d/d ratio, the bearing using oil grade 'A_1' had a higher load-carrying capacity per unit area than when using oil grades 'B_1' and 'C_1'. However, an increase in C_d/d ratio reduced the load-carrying capacity. The oil temperature in load-carrying oil film was also higher in the bearing when using oil grade 'A_1' than 'B_1' and 'C_1' (Fig. 7.6 and 7.7). That is why the load on the bearing must be judicially selected, considering the temperature rise in the bearing. At 3000 rpm, the load-carrying capacity of a bearing using oil grade 'A_1' reduced from 36.06 to 17.22 kg cm^{-2} (Fig. 7.8) by increasing the clearance ratio from 0.000 92 to 0.002 46. Similarly, under the identical change of clearance ratio, the load-carrying capacity of bearing reduced from 34.5 to 15 kg cm^{-2} (Fig. 7.9) and 26.6 to 12.5 kg cm^{-2} (Fig. 7.10), for bearings using oil grades 'B_1' and 'C_1', respectively.

For reliable optimum bearing operation at identical speed and load-carrying capacity, it is evident that bearings using oil grade 'A_1' operate with a lower viscosity integral and need higher clearance ratio than when oil grades 'B_1' and 'C_1' (Fig. 7.2–7.4 and 7.8–7.10) are used.

7.10.4 Comparison between analytical load-carrying capacity and experimental data

The experimental data on load-carrying capacity versus eccentricity ratio of the bearing ($C_d/d = 0.002\,90$ and 0.003 04, and using a lubricant with viscosity of 27.7 cP at 40 °C) operating at 2000 rpm, as reported by Ferron *et al.*,[10] was compared with the analytical data derived from the theoretical procedure given in this chapter, for a bearing having the same parameters (C_d/d = 0.0030, $l = 80$ mm, $d = 100$ mm) and using the identical lubricant 'C_1'. Load on the bearing at different eccentricity ratios was calculated using Fig. 7.4 and Equations (7.3), (7.13) and (7.14). The load-carrying capacity (P), thus calculated, is shown, along with the experimental values, in Fig. 7.12. At the eccentricity ratio of 0.75, the values of both analytical and experimental load-carrying capacity were found to be identical. However, in the operating range, i.e. in the range of eccentricity ratios varying from 0.6 to 0.75, the load-carrying capacity as per the developed analytical procedure matched the experimental values more closely than the theoretical values reported in Reference 10, as evident from Fig. 7.12.

7.11 Conclusions and recommendations

Based on the analysis given in this chapter, the following conclusions are drawn:[11,12]

- The procedure discussed in this chapter can be used to study the effect of oil grades and clearance ratios on the reliability of the performance of cylindrical bearings without going into detailed thermal calculations.
- The procedure facilitates determination of the optimum value of clearance ratio for bearings using different grades of oil at various loads and operating speeds. A change in the safe load-carrying capacity can be evaluated if the

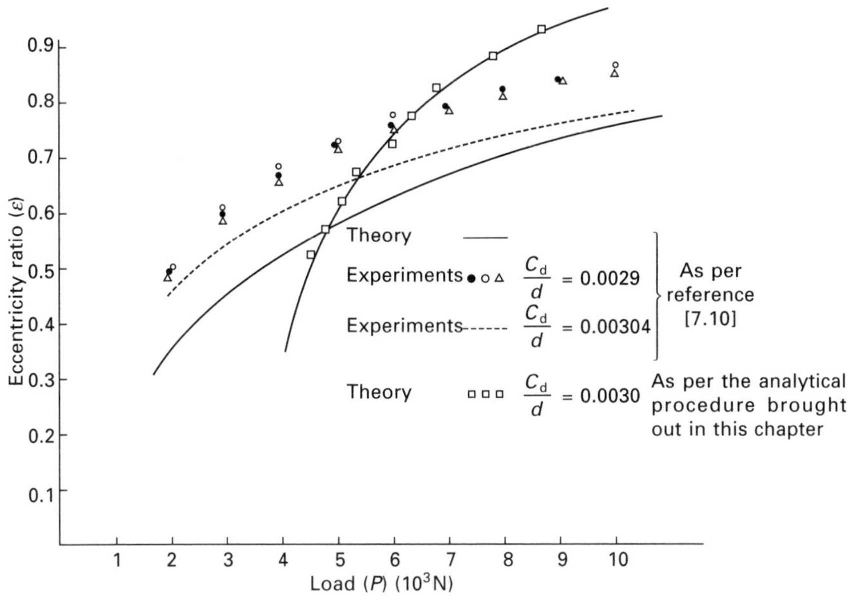

7.12 Eccentricity ratio versus load at operating speed of 2000 rpm.

bearing clearances are found to have changed at site or at the manufacturing stage.

- The optimum value of clearance ratio is much less for lighter grades of oil, in comparison with the heavier grades, for various operating conditions.
- The change in optimum clearance ratio is relatively negligible for viscosity integral above 9.7. However, with an increase in difference between viscosity integrals, the difference between optimum clearance ratios also increases under identical operating conditions.
- The increase in temperature in the load-carrying oil film is more in the case of the heavier grades of oil than the lighter grades under identical operating and parametric conditions. However, an increase in eccentricity ratio at a constant speed gives rise to higher temperature than an increase in speed at a constant eccentricity ratio, for bearings using different grades of oil.
- With an increase in clearance ratio, the safe load-carrying capacity per unit area reduces. However, for identical clearance ratio, a bearing using oil grade 'A_1' exhibits higher load-carrying capacity and temperature rise than when using oil grades 'B_1' and 'C_1'. But for identical load-carrying capacity and temperature rise, a bearing using oil grade 'A_1' needs a higher clearance ratio than when using oil grades 'B_1' and 'C_1'.
- Turbulence in the bearing may be correlated with viscosity integral, clearance ratio and other bearing parameters.

The reasonable agreement between the developed theoretical results and experimental results[10] validates the procedure developed in this chapter for assessment of the reliability of performance of cylindrical hydrodynamic bearings.

7.12 References

1 Ramsden, P., 'Review of Published Data and Their Application to the Design of Large Bearings for Steam Turbines', *Proc. Inst. Mech. Eng.*, **182**, pt 3A, 1967–68.

2 'Calculation Methods for Steadily Loaded Pressure-fed Hydrodynamic Journal Bearings', *Engineering Science Data*, Item No. 66023, ASME, 1965.

3 Yakhin, Z.A., 'Determination of Oil Temperature in Load Carrying Oil Film in Journal Bearings', *Russian Eng. J.* YDK 621, 436–233, 21.00.1.5, 21–23, 1973.

4 Nika, A., 'Friction and Thermal Characteristics of Radial Bearing', *Russian J. Friction Lubr.*, No. 3, Series F, 1–7, 1970.

5 Booser, E.R., Ryan, F.D. and Linkinhaker, C.L., '*Maximum Temperature for Hydrodynamic Bearings under Steady Load*', ASLE, 1970.

6 Martin, F.A. and Garner, D.R., 'Plain Journal Bearings under Steady Loads, Design Guidance for Safe Operation', Paper No. C313173, *European Tribology Conference*, 1972.

7 Petrunichev, A.E., 'Boundary Limit of Load-carrying Oil Film in Sliding Bearing', *Izveschiya AH USSR, Mashinovegenue*, 4, 24–30, 1971.

8 Contantinescu, V.N., 'Analysis of Bearings Operating in Turbulent Regime', *J. Basic Eng.*, 82–92, March 1982.

9 Wilcock, D.F., 'Turbulent Lubrication – Its Genesis and Role in Modern Design', *Trans. ASME*, January 1974.

10 Ferron, J., Frene, J. and Boncompain, R., 'A Study of the Thermohydrodynamic Performance of a Plain Journal Bearing – Comparison between Theory and Experiments', *ASME/ASLE Joint Lubrication Conference*, Washington, DC, 5–7, October 1982, Paper No. 82-Lub-16.

11 Prashad, H., 'The Effect of Viscosity and Clearance on the Performance of Hydrodynamic Journal Bearings', *STLE Trans.*, **31**(1), 113–119, 1988.

12 Prashad, H., 'A Simplified Procedure to Study the Effect of oil Grades and Clearance Ratios on Reliability of Performance of Cylindrical Bearings', *BHEL J.*, **10**(2), 39–56, 1989.

7.13 Nomenclature

A heat equivalent of work
C, D viscosity coefficients
C_d diameter clearance
C_v specific heat capacity of lubricant
d shaft diameter
l bearing length
L length of load-carrying oil film
n rpm

N	transition speed
p	load per unit area
P	load
R_{crt}	critical value of Reynolds number
R_e	Reynolds number
$\overline{S}$	loading coefficient
T	oil temperature
T_1	load-carrying oil film inlet temperature
T_2	load-carrying oil outlet temperature
T_e	effective oil temperature
T_{max}	maximum oil temperature in oil film
U	shaft rotational speed
V_x	oil velocity in *x*-direction
ε	eccentricity ratio
λ	thermal conductivity
μ	dynamic oil viscosity
μ_e	effective oil viscosity
ξ	oil film coordinate
ξ_1, ξ_2	non-dimensional coordinates of oil film
ρ	oil density
ν	kinematic oil viscosity
ψ	clearance ratio (C_d/d)

8
Spherical seating of hydrodynamic journal bearings

8.1 A general review

This chapter deals with a simplified methodology for the determination of optimum values of design parameters of spherical seating of hydrodynamic journal bearings. This methodology is developed on the basis of the minimum value of moment of friction and the friction moment constant in the spherical seating.

The optimum value for the radius of the spherical seating is determined from the inner and outer radii of the bearing and the angles formed by them, at the centre of spherical seating, with the axis of rotation. The optimum value of the axial length of the spherical seating is determined as a fraction of the bearing length, by equating the transverse load on the projected area of the bearing to that of the spherical seating. Nomographs are plotted for easy evaluation of the optimum values of the radius and axial length of spherical seating, by using the ratio of the outer diameter to the inner diameter, length and inner diameter of the bearing. Also, it is shown that an increase in the ratio of outer diameter to inner diameter, with a constant inner diameter, increases the minimum value of friction moment constant as well as the radius of the spherical seating, but reduces its axial length.

The design methodology given in this chapter, in addition to giving an approach for the determination of the optimum values of design parameters of spherical seating, acts as an effective tool to predict the performance of spherical seating of a hydrodynamic journal bearing.

8.2 Introduction

Spherical seating is generally provided to a hydrodynamic journal bearing with length-to-diameter ratio (L/D) of more than 1. The primary function of spherical seating is to facilitate bearing alignment during erection and rotation. In service, however, owing to frictional restraint, a motor bearing is unlikely to move in its spherical seat under shaft vibration or stator distortion of small

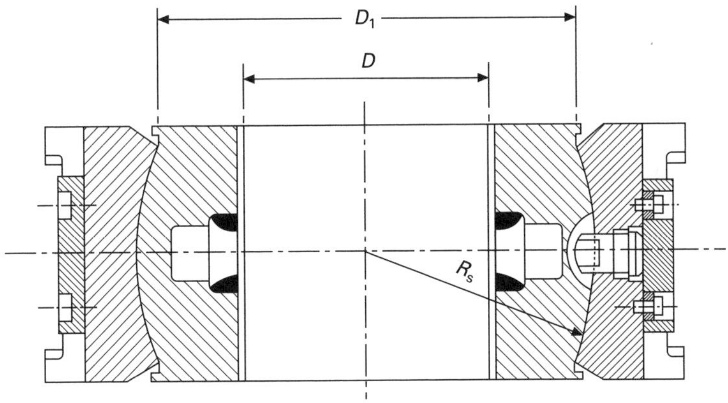

8.1 Spherical seating of hydrodynamic journal bearing.

amplitude. On the other hand, movement may take place in the spherical seat, particularly in a bearing with $L/D > 1$, if there is large angular displacement of the journal relative to the bearing pedestal. This ensures proper centring and self-alignment of a bearing relative to its housing.

Published literature for determination of dimensions of spherical seating is very scarce. However, a few papers dealing with the moment of friction and dimensions of a spherically seated bearing have been published,[1,2] but an easy design guidance for determination of the optimum values of radius and axial length of a spherical seating does not seem to have been developed. Optimum values of these dimensions would ensure bearing alignment with minimum values of moment of friction and friction moment constant.

In general, spherical seating to a cylindrical journal bearing is provided in two halves with similar spherical surfaces (Fig. 8.1). For preventing rotation of the bearing, stopper pins are used. Also, for controlling displacement, the spherical seating is generally designed with an interference fit.[3]

8.3 Theoretical basis of the simplified design methodology

A spherically seated bearing provides a self-aligning combined hydrodynamic journal and bi-directional thrust bearing (Fig. 8.1). Under the action of an applied force, the rotating system is subjected to a moment of friction. The magnitude of this moment is generally less than the bending moment in the spherical seating, otherwise the bearing itself may rotate in its spherical seating. However, for a quick and sensitive response of a bearing system, the optimum value of the radius of spherical seating is worked out on the basis of the minimum bending moment so as to ensure self-alignment in the spherical seating and a proper centring for a minimum friction moment in the bearing.

Generally, in a rotating system, the component of a force, in the direction perpendicular to the x and z directions, is minimum, i.e. $P_y = 0$ (Fig. 8.2). Hence, the moment of friction M_x at the spherical seating due to the P_z component of force, in a bearing with inner radius R, is given as:

$$M_x = P_z R f F_m \qquad [8.1]$$

where F_m the constant of the moment of friction, depends on the coefficient of friction between the spherical surfaces (f) and various other coefficients depending on the values of angles γ and γ_1.[1] From Fig. 8.2, it is evident that:

$$\sin \gamma = R/R_s \qquad [8.2]$$

and

$$\sin \gamma_1 = R_1/R_s \qquad [8.3]$$

It follows from the above equations that:

$$\gamma_1 = \arcsin (D_0 \sin \gamma) \qquad [8.4]$$

and

$$\gamma = \arcsin (\sin \gamma_1/D_0) \qquad [8.5]$$

where

$$D_0 = R_1/R \qquad [8.6]$$

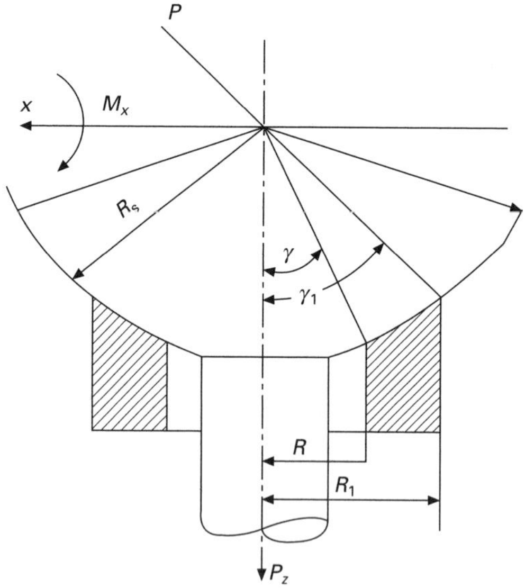

8.2 Design parameters of a spherical seating.

8.4 Evaluation of minimum values of constant of moment of friction and optimum values of the design parameters of spherical seating

8.4.1 Constant of moment of friction

The moment of friction in a bearing depends on the component of applied force, inner radius and coefficient of friction F_m. Theoretical determination of F_m is an unresolved and critical problem. However, F_m has been determined experimentally for $f = 0.2$ at different values of angles γ_1, γ and D_0.[2] It is shown that F_m has a minimum value for different values of D_0, depending on the optimum values of angle γ_1. The minimum value of F_m for different values of D_0 and optimum values of γ_1 are given in Table 8.1.[2]

8.4.2 Outer radius of bearing

In bearing design, the inner radius R and bearing length L are usually known. However, these parameters can also be determined from the radius of the journal.[4,5] The optimum value of outer radius of bearing R_1 can be calculated from R and the allowable safe unit average pressure (p) exerted by P_z, the component of the applied force on the projection of the bearing surface. It is given as:

$$p = \frac{P_z}{\pi(R_1^2 - R^2)} \quad [8.7]$$

For reliable operation, p must always be less than p_{max}, which is the maximum allowable safe unit pressure on the bearing material. Hence, if p_{max} is known, then R_1, the optimum value of the outer radius of a bearing with inner radius R, under the maximum design limit of applied force, can be determined. The

Table 8.1 Minimum values of constant of moment of friction F_m at various optimum values of D_0 and angle γ_1

Serial no.	Ratio of outer diameter to inner diameter of bearing (D_0)	Optimum value of angle formed by outer radius with bearing axis at centre of spherical seating (γ_1)	Minimum value of constant of moment of friction (F_m)
1.	1.1	54°	1.65
2.	1.2	57°	1.75
3.	1.3	60°	1.87
4.	1.4	63°	1.95
5.	1.5	65°	2.05

ratio of the outer diameter to the inner diameter (D_0) can then be determined by using Equation (8.6).

8.4.3 Radius of spherical seating

On calculating D, the optimum value of the angle formed by the outer radius with the axis of the bearing at the centre of spherical seating γ_1 is determined for the minimum value of moment of friction constant F_m, with the help of Table 8.1. The value of γ is calculated from the value of D_0 and angle γ_1 thus determined using Equation (8.5). The optimum value of radius of spherical seating R_s is calculated by using Equation (8.3).

8.4.4 Axial length of spherical seating

The axial length of the spherical seating is evaluated from load criterion of a bearing. The transverse load on the projected area of a hydrodynamic journal bearing is equated to that on the spherical seating. Since the length of a bearing (L) is known, the axial length of the spherical (L_s) can be calculated as a fraction of L by using the following equations:

$$P = pLD = p\, L_s D_s \qquad [8.8]$$

Hence

$$L_s = LD/D_s \qquad [8.9]$$

8.5 Functional nomographs for evaluation of optimum values for spherical seating design parameters

The values of D_1, R_s, L_s and angle γ were determined for different values of D_0 and D varying between 1.1 and 1.5 and from 50 to 400 mm, respectively, by using Equations (8.2), (8.5), (8.6) and (8.9), for optimum values of the angles γ_1 (Table 8.1).

For an easy and quick evaluation of the optimum dimensions of the spherical seating of a bearing, nomographs (Fig. 8.3 and 8.4) were plotted. Figure 8.3 was used to determine the values of D_0 from the predetermined values D and D_1, in order to evaluate the optimum values of the angles γ_1 and γ. Figure 8.4 was used to determine the optimum values of the radius R_s and the axial length L_s of spherical seating, from the predetermined values of D_0 and D. The axial length of the spherical seating was expressed as a fraction of the length of the bearing (Fig. 8.4). The procedure to use the nomographs for determination of optimum values of angles γ, γ_1 and the dimensions R_s, L_s is illustrated in Fig. 8.3 and 8.4.

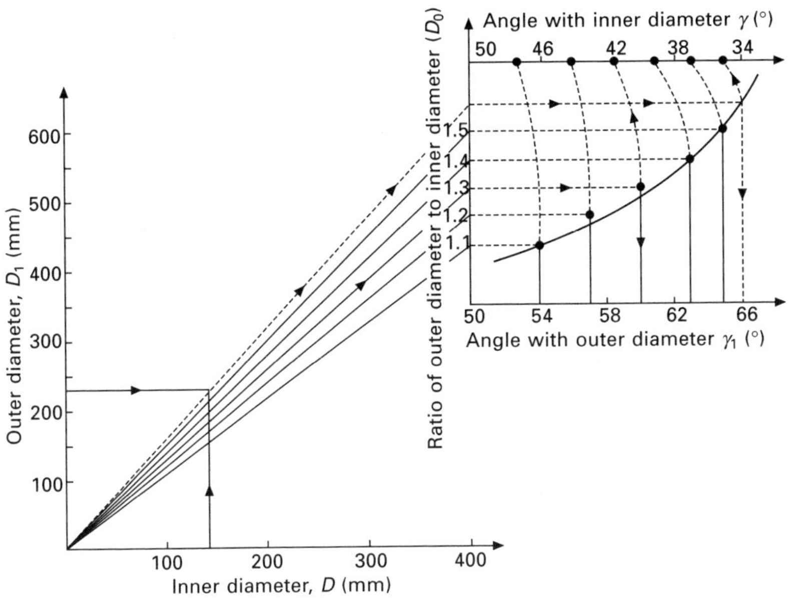

8.3 Evaluation of variation of optimum values of angles γ and γ_1 with various ratios of outer diameter to inner diameter (D_0) of a journal bearing.

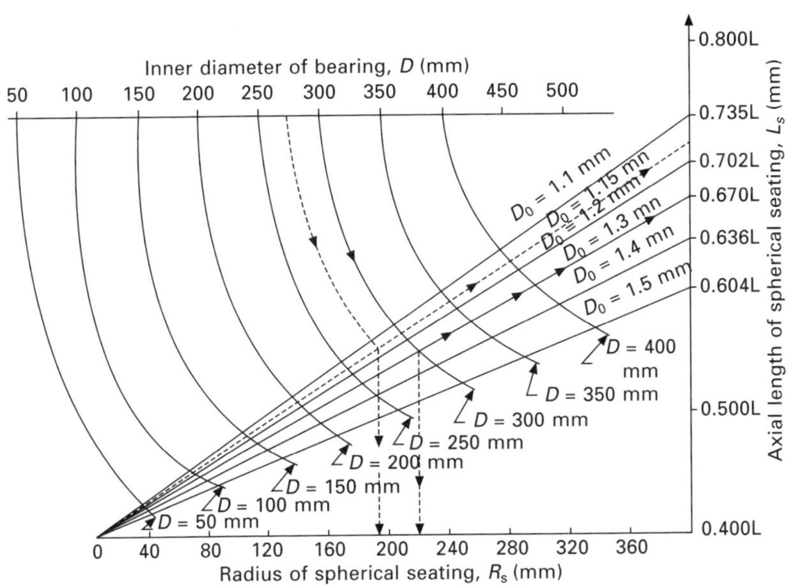

8.4 Evaluation of optimum values of radius (R_s) and axial length (L_s) of spherical seating for various values of inner diameter (D) and ratios of outer diameter to inner diameter (D_0) of a journal bearing.

8.6 Guidelines for choosing the optimum values for spherical seating design parameters

8.6.1 Ratio of outer diameter to inner diameter

As the ratio (D_0) of outer diameter to inner diameter of the bearing increases from 1.1 to 1.5, the minimum value of the friction moment constant varies from 1.65 to 2.05 with the optimum value of angle γ_1 varying from 54° to 65° (Table 8.1). This shows that to keep the values of F_m and the moment of friction to a minimum (Equation 8.1), it is necessary to restrict D_0 to a minimum. Thus, a bearing should be designed judiciously with minimum D_1, taking into account the required safe load-carrying capacity. However, considering bending and shear deformation, the optimum value of D_0 reduces for a bearing with higher L/D ratio (approximately 1.26 for $L/D = 2$ as against 1.625 for $L/D = 0.5$). The optimum value of D_0 is 1.5, considering shear deformation at different L/D ratios.[6]

8.6.2 Radius of spherical seating

The optimum value of the radius of spherical seating (R_s) depends on D_0 and the inner diameter (D) of the bearing. Increase in D_0 at a given D increases the radius of spherical seating (R_s). For a bearing with inner diameter of 300 mm, the optimum value of (R_s) increases from 203.70 to 248.25 as the value of D_0 changes from 1.1 to 1.5 (Fig. 8.4). To keep the moment of friction to a minimum, it is necessary to reduce D_0 and hence, R_s.

8.6.3 Axial length of spherical seating

The optimum value of the axial length of spherical seating (L_s) depends on its radius (R_s), inner diameter (D) and, in turn, D_0 of the bearing (Equations 8.6 and 8.9). The higher the value of D_0, the lower will be the optimum value of the axial length of spherical seating (Fig. 8.4). On varying D_0 from 1.1 to 1.5, the optimum value of L_s changes from 0.735 to 0.604 times the length of the bearing (L), as shown in Fig. 8.4.

8.7 Conclusions and recommendations

The main conclusions and recommendations are as follows:[7]

- The simplified design methodology given in this chapter can be used to determine the optimum values of design parameters of spherical seating of a hydrodynamic journal bearing, without detailed design calculations.
- To keep the value of friction moment constant to a minimum, the ratio of outer diameter to inner diameter of a bearing should also be kept to a

minimum. However, the value of the outer diameter is to be judiciously chosen by taking into account the required safe load-carrying capacity of the bearing.

- Increase in ratio of outer diameter to inner diameter, for a given value of the inner diameter of the bearing, increases the optimum value of radius of the spherical seating but reduces the optimum value of its axial length.

The accuracy of the simplified design methodology depends on the variation of coefficient of friction between the spherical surfaces and, in turn, the change in moment of friction moment constant of the bearing. The optimum values of design parameters of spherically seated bearings, as calculated by this design methodology, have been found to closely match those arrived at by detailed calculations.

8.8 References

1 Zandfos, L.V., 'Determination of Dimensions of a Spherical Bearing', *Vestnik Mashinostroeniya*, **56**(7) (V.D.C. 621.828.001.24 *Russian Eng. J.*), 37–38, 1976.

2 Zandfos, L.V., 'Moment of Friction of a Spherical Bearing', *Izv. Vuzov, Mashinostroenie*, 9, 34–38, 1973.

3 Neale, N.J., '*Tribology Hand Book*', Newnes, Butterworth, 1970.

4 Prashad, H., 'The Effects of Viscosity and Clearance on the Performance of Hydrodynamic Journal Bearings', *ASLE 42nd Annual Meeting*, Anaheim, California, 11–14, May 1987. Also published in *STLE Trans.*, **31**(1), 113–119, 1988.

5 Martin, F.A. and Garner, D.R., 'Plain Journal Bearings under Steady Loads – Design Guidance for Safe Operation', Paper No. C313/73, *First European Tribology Conference*, pp. 449–463, 1973.

6 Peeken, H., 'The Effect of Bearing Design on the Reliability of Journal Bearing Behaviour', Institute of Mechanical Engineers, Paper No. C 308/73, pp. 407–413, 1973.

7 Prashad, H., 'A Simplified Design Methodology for Determination of Optimum Values of Design Parameters of Spherical Seating of Hydrodynamic Journal Bearings", *BHEL J.*, **10**(1), 45–50, 1989.

8.9 Nomenclature

D	bearing inner diameter
D_1	bearing outer diameter
D_0 (= D_1/D)	ratio of outer diameter to inner diameter
D_s	diameter of spherical seating
f	coefficient of friction between spherical surfaces
F_m	constant of moment of friction
L	bearing length
L_s	axial length of spherical seating
M_x	moment of friction

p	safe unit average pressure
p_{max}	maximum safe unit pressure
P	force acting on bearing
P_x	component of force in x-direction
P_y	component of force in y-direction
P_z	component of force in z-direction
R	bearing inner radius
R_1	bearing outer radius
R_s	radius of spherical seating
γ	angle formed by the bearing inner radius with the axis of rotation at the centre of spherical seating (in degrees)
γ_1	angle formed by the bearing outer radius with the axis of rotation at the centre of spherical seating (in degrees)

9

Life estimation of turbine oils: a methodology and criterion for acceptance or rejection

9.1 A general review

The function of turbine oil is to reduce friction and wear in bearings, to serve as a coolant and sealant, to act as the hydraulic medium in the governor and to protect metallic surfaces from rusting and corrosion. Oils have a tendency to become oxidized during usage especially in the presence of heat, water, air and certain metallic impurities, which act as catalysts for oxidation.

Oxidation of oil causes formation of acid products, resulting in the increase of total acid number (TAN) and decrease of rotating bomb oxidation test (RBOT) values. Oil manufacturers have correlated RBOT, percentage depletion of antioxidant and acid values with turbine oil stability test (TOST) values and proposed a certain definite RBOT value as rejection criterion for turbine oil. However, there is no universally accepted methodology and criterion for acceptance.

In this study, RBOT values of various turbine oils have been determined for fresh oils after TOST of different durations in the laboratory and for the used oils received from different sites. Also, an attempt is made to establish a methodology to correlate kinetics of oxidation of oils using RBOT and TOST values. The quality of oils, as well as acceptance/rejection criterion to determine life of the turbine oil, based on the experimental results, are discussed.

9.2 Introduction

Modern turbine oil should have a very high level of functional properties and the oil has to be formulated such that these properties are maintained at an adequately high level during usage in the system. Furthermore, turbine oil should have good demulsibility, viscosity and resistance to air entrainment and foaming as well as anticorrosion and rust protection characteristics. Oils have a tendency to become oxidized during usage. When oil undergoes oxidation, its ability to shed water reduces and a permanent water emulsion is formed. Emulsified oil cannot provide adequate oil films in bearings and in extreme cases causes scoring on the surfaces of bearings and gear teeth. Also, excessive foaming can interfere with the heat-removing capability and promote further oxidation.

Oxidation of oil causes formation of acid products resulting in the increase of TAN and decrease of RBOT values.[1] TOST and RBOT values were used to establish the quality of an oil as well as acceptance/rejection criterion to determine the residual life of turbine oils.[2,3]

A kinetic approach was also adopted to arrive at a comparative evaluation of turbine oils. The oils were degraded in the laboratory in RBOT apparatus at different temperatures, followed by RBOT life estimation, as per ASTM-D 2272.

Laboratory tests of fresh turbine oils ('A', 'B' and 'C') from three different suppliers were carried out before and after ageing under TOST conditions of different durations. Under both these conditions, the measurement of properties of RBOT, TAN and EA (elemental analysis) were carried out to establish the deterioration of the oils. The correlations of TOST, TAN, RBOT values and EA have been used for analysis to evaluate the quality of the turbine oils investigated.

9.3 Experimental investigations

9.3.1 Turbine oil stability test (TOST)

For TOST, the ASTM-D 943 method was used. TOST ageing was carried out for oils 'A', 'B' and 'C' received from different suppliers. TOST ageing was carried out up to 2000 h and changes in properties such as TAN, RBOT and EA were established to study the severity of the deterioration. TOST basically signifies the oxidation stability of turbine oils using different crudes, effect of processing, blending and additive packages.

9.3.2 Rotating bomb oxidation test (RBOT)

RBOT ageing of fresh oil samples ('A', 'B', 'C') from different suppliers was carried out at 140, 150 and 160 °C for kinetic studies (as per ASTM-2272). A plot of log of RBOT value versus inverse of temperature ($1/T$) in degree absolute was generated (Fig. 9.1).

Activation energies were calculated from the slope of the plot by the Arrhenius equation. Furthermore, RBOT life was derived from the graph at 90 °C for all three samples. Thus, RBOT measurements were carried out on fresh, aged and service oils received from different sites.

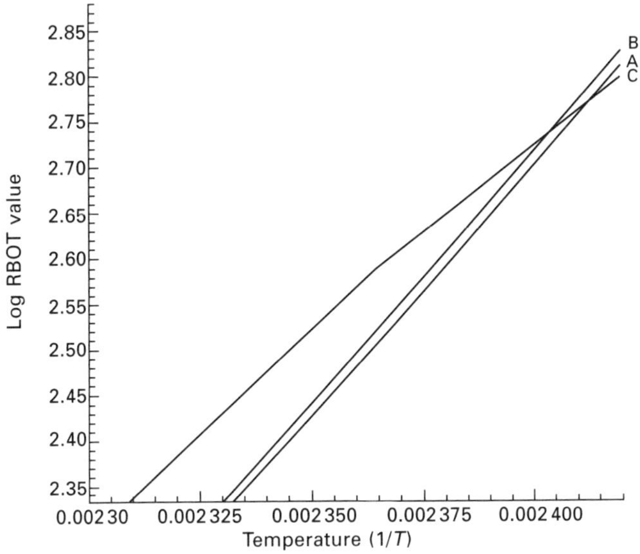

9.1 Variation of the log of RBOT values versus the inverse of absolute temperature for various oils.

9.3.3 Four-ball tests

As these turbine oils are recommended to be used for geared turbines, the oils are expected to possess load-carrying capacity, which is measured by FZG. To establish these characteristics and also to distinguish between the characteristics of extreme pressure and anti-wear additives, four-ball EP (extreme pressure) and four-ball wear tests, as specified in IS 8406 and revised in 1993, were carried out.

9.4 Data deduction

General properties of the laboratory tests of fresh oil samples 'A','B' and 'C' have been determined as per ASTM and found to be within the specified limits (Table 9.1). Data diagnosed by RBOT and TAN values of fresh oils and after TOST ageing for 1500 and 2000 h (as per ASTM-D943) are given in Table 9.2. Table 9.3 gives data of RBOT at various temperatures (140, 150, and 160 °C) for all the fresh oils. Behaviour and service life of each oil, supplier-wise, were determined and are shown in Table 9.4. Data of activation energies, as calculated from log time versus inverse of temperature as depicted in Fig. 9.1 (kinetic studies), is shown in Table 9.5 along with RBOT life of turbine oils at 90 °C.

Table 9.1 Properties of turbine oils

Sl no.	Property	Oil 'A'	Oil 'B'	Oil 'C'
1.	Density at 15 °C	0.870	0.865	0.868
2.	KV at 40 °C KV at 100 °C	48.140 7.100	46.400 7.120	46.500 7.500
3.	Viscosity index	110	110	110
4.	TAN	0.1	0.1	0.04
5.	Pour-point	–6	–6	–6
6.	Demulsibility			
(a)	Free water (ml) collected during the test (starting with 45 ml)	41	41	39
(b)	% water in oil after 5 h test	0.35	0.3	0.4
(c)	% left after centrifuging	Nil	Nil	Nil
7.	Foaming characteristics			
	at 24 °C	Nil	Nil	Nil
	at 93.5 °C	Nil	Nil	Nil
	at 24 °C after the test at 93.5 °C	Nil	Nil	Nil
8.	Air release value at 50 °C	1 min 8 s	1 min 8s	1min 20 s
9.	Four ball EP test			
(a)	Weld load (kg)	150	150	150
(b)	Mean Hertz load (MHL) (kg)	39	46	43

9.5 Results and discussion

9.5.1 Analysis and life estimation of oil 'A'

RBOT values of TOST ageing as per ASTM D 943 of fresh oil 'A' after 2000 h and 1500 h as determined experimentally, are 140 and 157 min, respectively. The higher RBOT value of 157 min at 1500 h indicates that the oil got less oxidized than at 2000 h having RBOT of 140 min. The change in RBOT values for fresh oil from 317 min to 157 min after 1500 h of TOST shows rapid oxidation of the oil compared with that from 1500 h to 2000 h where the change in RBOT value is only from 157 to 140 min. This indicates initial fast ageing followed by slow ageing. The initial ageing of 53% by RBOT in 1500 h reduces to 5% in this span of 1500 h to 2000 h of use. The degradation of the oil is also confirmed by the increase in the values of iron and copper as determined by elemental analysis (Table 9.2).

Table 9.2 RBOT and TAN values, and elemental analysis levels of fresh turbine oils and after TOST ageing for 1500 and 2000 hours (as per ASTM-D943)

	Oil 'C'			Oil 'B'			Oil 'A'		
	TAN (mg g^{-1}KOH)	RBOT (min)	E A ppm (Fe, Cu, Zn)	TAN (mg g^{-1}KOH)	RBOT (min)	EA ppm (Fe, Cu, Zn)	TAN (mg g^{-1}KOH)	RBOT (min)	EA ppm (Fe, Cu, Zn)
Fresh oil	0.04	385	35, <2, <2	0.1	327	38, <2, <2	0.1	317	28, <2, <2
Ageing for 1500 h	0.16	210	36, <2, <2	0.23	220	96, 18, <2	0.3	157	110, 100, 2
Ageing for 2000 h	–	72	–	–	172	–	–	140	–

Table 9.3 Kinetic study of RBOT life of fresh turbine oils at different temperatures (as per ASTM-D2272)

Oil	RBOT values (min)		
	140 °C	150 °C	160 °C
'C'	660	385	215
'B'	680	327	168
'A'	655	317	165

Table 9.4 Service life comparison of turbine oils

Turbine oil used in the machine	Site	Fresh oil RBOT (min)	The 25% cut-off value (min)	RBOT of TOST aged oil after 2000 h (min)	Service oil RBOT (min)	Machine running time at the time of sample withdrawal (years)
'C'	Z	385	96	72	340	3
'B'	Y	327	82	172	245	5
'A'	X	317	79	140	215	2

Table 9.5 RBOT life and activation energy of turbine oils

Oil	Activation energy ΔE (kcal)	RBOT Life of oil at working temperature of 90 °C (min)
'C'	21.6	18750
'B'	24.6	41336
'A'	24.5	39738

Ratio of RBOT life of oil 'B' to oil 'A' is 1.04.
Ratio of RBOT life of oil 'B' to oil 'C' is 2.20

Fresh oil 'A' indicates an RBOT value of 317 min at 150 °C (Tables 9.2 and 9.4). Its 25% cut-off value is 79 min, which is generally considered to be the oil reclamation/rejection limit as per ASTM-D 4378 (Table 9.4).

The RBOT value reaches the cut-off limits of 79 min after 2650 h of TOST ageing as per graphic evaluation. Also, oil 'A' sample collected from site 'X' after 2 years of service showed an RBOT value of 215 min (Table 9.4). This is well above the 25% cut-off value. This shows that the oil has not reached its reclamation or its rejection limits as per RBOT property. From the RBOT values of aged oil samples taken from service intermittently from site 'X', it is estimated that the 25% cut-off value of 79 min will be reached approximately after 6 years (extrapolated graphical estimation).

Thus 2650h of TOST ageing = 6 years of residual service life of 'A' oil approximately. Thus, the net life of fresh oil 'A' is estimated as 8 years.

9.5.2 Analysis and life estimation of oil 'B'

Fresh oil 'B' indicates an RBOT value of 327 min at 150 °C (Table 9.3). Its 25% cut-off value is 82 min (rejection limit as per ASTM).

RBOT value of TOST ageing of fresh oil 'B' after 2000h was determined as 172 min (Tables 9.2 and 9.4), which is above the rejection limit. RBOT value reached the cut-off limit of 82 min after 3450h of TOST ageing (graphical value). Furthermore, the higher RBOT value of 220 min at 1500h as determined experimentally indicates that the oil is gradually oxidized from its RBOT value of 327 min (fresh oil) to 172 min (after 2000h). This can be assessed from the fact that up to 1500h of TOST ageing, RBOT is reduced by 33%, followed by 15% in the span of 1500 to 2000h. The change of ppm levels in EA indicates the deterioration of the oil (Table 9.2).

Also, oil 'B' sample collected from site 'Y' after 5 years of service showed an RBOT value of 245 min (Table 9.4). This value is well above the 25% cut-off limit and it shows that the oil had not reached its reclamation/rejection limit with respect to RBOT property. From the RBOT values of aged oil samples taken from service intermittently from the site 'Y', it has been estimated that the 25% cut-off value of 82 min will be reached approximately after 10 years (extrapolated graphical value).

Thus 3450h of TOST ageing = 10 years of residual service life of oil 'B' approximately. So, the net life of oil fresh 'B' is estimated as 15 years.

9.5.3 Analysis and life estimation of oil 'C'

The RBOT value of fresh oil 'C' was determined as 385 min at 150 °C as shown in Tables 9.2 and 9.4. Its 25% cut-off value is 96 min.

The RBOT value of TOST ageing of fresh oil 'C' after 2000h has been determined as 72 min (as shown in Table 9.2). The significant change in RBOT value of fresh oil, i.e. 385 min to 210 min after 1500h and 72 min at 2000h, shows that oil 'C' deteriorates and gets oxidized at a higher pace with the higher duration of use. This can be assessed from the fact that after up to 1500h of TOST ageing, RBOT is reduced by 45% followed by 36% in the span of 1500h to 2000h. This shows that it may not give a longer life, but for medium life this oil is adequate. No significant change in ppm levels is indicated in the EA. This shows that the inhibitors used in the oil are quite stable up to 1500h of TOST. It has been determined graphically that RBOT value of oil 'C' reaches the cut-off limit of 95 min after 1850h of TOST ageing.

Also, oil 'C' sample collected from site 'Z' after 3 years of service has shown an RBOT value of 340 min (Table 9.4). This value was above 25% cut-off limit like oils 'A' and 'B', and it shows that the oil 'C' had not reached its reclamation/rejection limit with respect to RBOT property in this span of use. But for higher duration of use, it may deteriorate very fast as indicated in laboratory investigations of RBOT value of 72 min after 2000 h of TOST as against 210 min after 1500 h of TOST (Table 9.2). It is further confirmed from the RBOT value of the aged samples taken from service intermittently (from site Z). It has been estimated that the 25% cut-off value of 96 min will be reached approximately after 3 years (extrapolated graphical value).

Thus 1850 h of TOST ageing = 3 years of residual service life of oil 'C' approximately. So, the net life of fresh oil 'C' is estimated as 6 years.

9.5.4 Four-ball test results

It was observed that the results of weld load test of all the three oils did not conform to the IS specification No. 8406, 1993, because the quantity of EP additive used in the oils was not adequate. However, the mean Hertz load value was found to be within the range. Also, four-ball wear test indicated that the scar diameter is within the specified limit.

In addition to the above, the kinetic studies conducted at IIT, New Delhi, on the above three oils indicated the same order of the quality as found in investigations carried out at BHEL, Corporate R&D Division.

9.5.5 Overall comparison of the oils

Fresh oils 'A', 'B' and 'C' have almost identical characteristics and meet the general specifications. However, oils 'A' and 'B' have higher TAN values than oil 'C'. The TAN of fresh oil 'C' is 0.04 $mg\,g^{-1}$ KOH in comparison with 0.1 $mg\,g^{-1}$ KOH for oils 'A' and 'B' against the normal specified standard value of 0.2 $mg\,g^{-1}$ KOH. This shows that ratio of TAN value of oils 'A' and 'B' to that of oil 'C' is 2.5. This results in slow increase of TAN value of oil 'C' during TOST ageing up to 1500 h as indicated in Table 9.2. The ratios of TAN value of oil 'B' to oil 'C' and oil 'A' to oil 'C' after 1500 h of TOST ageing are 1.44 and 1.89, respectively, as against 2.5 for fresh samples of these oils. RBOT values of all the fresh oils were found to be more or less identical. The ratios of RBOT of fresh oils 'B' and 'C', and oils 'A' and 'C' vary between 0.86 and 0.83, respectively. After 1500 h of TOST, RBOT ratios of oils 'B' and 'C', and oils 'A' and 'C' change to 1.05 and 0.77, respectively. But after 2000 h of TOST, the deterioration of oil 'C' is very significant. The ratios of RBOT after 2000 h of TOST for oil 'B' and 'C', and for oils 'A' and 'C', change to 2.4 and 1.94, respectively, whereas this ratio for the oil 'B' to 'A' is 1.23. From this, it is evident that the performance of oil 'C' deteriorates much faster after long use as compared with oils 'A' and 'B', and the oil loses its inhibition oxidation properties. However, the lesser value of TAN makes oil 'C' stable against oxidation up to 1500h of TOST ageing (Table 9.2).

Similar conclusions can be drawn by the kinetic study of RBOT life of fresh turbine oils at different temperatures. At 140 and 150°C, there is an insignificant change in the RBOT values of these oils: the values change only from 655 to 680min at 140°C and 317 to 385min at 150 °C, but at 160°C RBOT values for oil 'C' are 215min as against 168min for oil 'B' and 165min for oil 'A' (Table 9.3). The higher difference in RBOT value with increase in temperature indicates that oil 'C' is deteriorating at a higher rate than 'A' and 'B' at higher temperature of use and will have a comparatively shorter life. From the activation energy, it is further confirmed that oil 'C' is inferior to that of oils 'A' and 'B' because activation energy of oil 'B' and 'A' are 24.6kcal and 24.5kcal, respectively, as against 21.6kcal for oil 'C' as shown in Table 9.5.

Furthermore, results of kinetic studies at 90°C of RBOT working temperature indicates that RBOT life of oil 'B' is 2.20 times that of oil 'C' and 1.04 times that of oil 'A' (Table 9.5).

9.6 Conclusions and recommendations

The following may be concluded based on the studies conducted:[4]

- The life of oil 'B' is 2.20 times that of oil 'C' and 1.04 times that of oil 'A'.
- The deterioration of oil 'B' is gradual and uniform with time as against that of oils 'A' and 'C'.
- The lesser TAN value of oil 'C' makes it stable for medium duration of use. For higher duration of use, the deterioration is very fast.
- A gradual and uniform deterioration of oil by RBOT not exceeding 33% and 15% may be allowed as a criterion after TOST ageing of 1500 and 2000h, respectively. This is required to achieve the life of 15 years of turbine oils in actual site use.
- The service life of the oils can be assessed from the TOST and RBOT values.
- Based on the TOST ageing up to 1500 and 2000h as monitored by RBOT, TAN and RBOT kinetic studies, the quality of the tested oils is of the following order: oil 'B' > oil 'A' > oil 'C'.

In short, turbine oil 'B' is found to be superior to oils 'A' and 'C'. This may be due to the blending of various additive packages. However, it may be necessary to monitor each batch of fresh oils for properties to assess the quality of the oils since batch-to-batch variation was observed in the studies.

9.7 References

1 Goel, P.K., Jaya Prakash, K.C and Srivastava S.P., 'Oxidation Stability of Steam Turbine Oil and Laboratory Methods of Evaluation', *ASLE Trans.*, **2**, 89–95, 1984.

2 Warne, T.M. and Vienna, P.C., 'High Temperature Oxidation Testing of Lubricating Oil', *ASLE Trans.*, **40**, 211–217, 1984.

3 Herderm, D., Vienna, P.C., 'Control of Turbine Oil Degradation During Use', *ASLE Trans.*, **37**, 67–71, 1980.

4 Murthy, T.S.R., Prashad, H. and Jagga, C.R., 'Life Estimation of Turbine Oils – A Methodology Criterion for Acceptance', *Petrotech 2003*, 9–12, January, New Delhi.

10
Axial force on motor bearings: a tool for performance evaluation

10.1 A general review

This evaluation deals with the investigations carried on the different AC horizontal motors to monitor their performance, and also to assess the axial force on the motor bearings. Under 'load' conditions, the axial force on the bearings is a net resultant of the forces generated by both driven assemblies, and it is affected by operating parameters, design and alignment of the system. However, under 'no-load' operation of a motor, the axial force experienced by the motor bearings reveals the design, manufacturing accuracy, constructional tolerances, asymmetry due to imperfect centring of the rotor in the stator box, non-linear stator bore, non-circular bore, bend shaft, shaft deflection and overhanging weight of motor. Thus, the axial force on the bearings under 'no-load' conditions reflects the 'quality' of a motor.

The investigations reveal that under 'no-load' the axial force experienced by various motor bearings of good quality motors is very low and subsequently will have no effect on the behaviour and operating life of the bearings. Also, it confirms that motors with low axial force on bearings have been manufactured to a reasonable accuracy. However, the axial force under load, exerted by the driven equipment, may affect the performance and life of the motor bearings.

10.2 Introduction

The axial force on the bearings of a rotor of a horizontal AC motor appears to be due to an unbalanced magnetic pull as soon as the motor is started. The unbalanced magnetic pull (UMP) in a motor is generated because of a variation in the air gap and eccentricity of a rotor. The relation between UMP and eccentricity of a rotor can be determined through the measurement of an electromagnetic field at the rotor surface. Also, the significance of the tangential flux component in an air gap can be assessed.

The radial magnetic force in a motor is in the direction of the smallest air gap and this tends to diminish the gap, i.e., increases the rotor deflection.

Therefore, there is a tendency for eccentricity to increase under the influence of radial magnetic force. However, the flexure stiffness of a shaft has a restoring effect. Besides this, under these conditions, a change of position of the axis of rotation of a shaft from the symmetrical position also takes place. This originates an axial force, which acts as a restoring force, and unlike the radial pull, it tends to reduce the asymmetry. If the rotor is both radially and axially asymmetrical, then the resulting force occurs at a certain angle between 0° and 90° with the axis of rotation. The resulting force can load the motor bearings beyond the permissible limit and can lead to an excessive temperature rise, vibrations and premature failure of the bearings.

Under 'load' conditions, the axial force on the bearings is a net resultant of the forces generated by both drive and driven assemblies, and is affected by operating parameters, and the design and alignment of the system. However, the axial force experienced by the motor bearings under 'no-load' reveals the design, manufacturing accuracy, constructional tolerances and asymmetry due to imperfect centring of the rotor in the stator box, non-linear stator bore, non-circular bore, bend shaft, shaft deflection, over-hanging weight of a motor. Thus, the axial force on the bearings under 'no-load' conditions reflects the 'quality' of a motor.

To diagnose the 'quality' of a motor, an 'axial force' measurement technique has been developed. Also, the performance of different horizontal motors has been assessed under normal and misaligned conditions on the test bed.

10.3 Axial force measurement technique

To monitor the axial force developed by the rotating shaft of a motor, the special strain gauge type load cell with a free rotating ball (30.33 mm in diameter), mounted inside the cavity of load cell, was used. The ball of the load cell, which just touched the motor shaft end, rotated as soon as the motor is switched on. The axial force experienced by the ball was transmitted to the strain gauge bridge of the load cell and then to the millivolt digital indicator. The instrumentation scheme, shown in Fig. 10.1, consisted of load cell, model CB-8, serial 238 (Techno-lab) and 4.5 digit Fluke digital multimeter model 8050 A. The power supply unit gave 10 V (DC) excitation voltage to the load cell. The load cell and Fluke digital multimeter were calibrated precisely on the computerized controlled universal load testing machine with gradual ascending and descending loads, and millivolt outputs were programmed. The average mV output was correlated with the load in N. Thus, the relation mV/N = 1.09×10^3 was established and used during the test for monitoring the axial force experienced by the bearings.

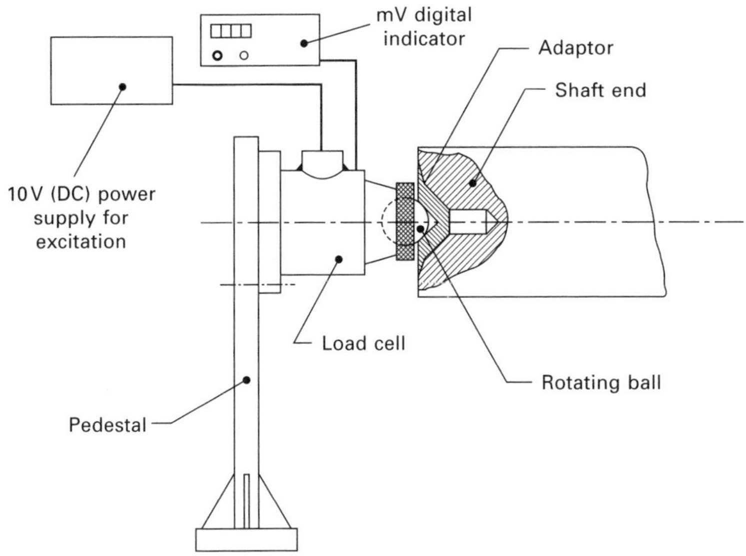

10.1 Instrumentation scheme for measurement of axial force on the motor bearings.

10.3.1 Accessories for measurement

The pedestal for holding the load cell and keeping it at variable heights – in line with the axis of rotation of the motors – had been specially designed so that the same pedestal could be used for various motors of different pedestal heights. An adaptor was also developed and used between the ball of the load cell and a motor shaft for better contact of the ball with a shaft during rotation of the rotor of the motor. An adequate amount of lubricant was applied at the contact point between the ball and the adaptor, fitted at the shaft end, to reduce frictional losses.

10.3.2 Procedure for axial force measurement

The load cell was firmly mounted on the pedestal, which was clamped rigidly to the base plate. Furthermore, the axes of rotation of the ball of the load cell and the rotor were precisely matched. The contact of the ball with the shaft end through the adaptor was adjusted such that the ball just touched the adaptor, and was able to rotate freely. Under this condition the digital multimeter indicated the reference load under stationary and rotating conditions, so the force experienced by the bearing was experimentally established. The direction of the axial movement of the shaft was assessed by the increase in the magnitude of the axial force compared with the reference load under stationary conditions, indicated by the digital indicator. The direction of the axial movement was also confirmed by monitoring the axial force on the other free end of the motor shaft.

10.4 Experimental determination of axial force

The rotor of a horizontal AC motor is generally supported by a roller bearing at the floating end to take the radial load and by the ball bearing at the locating end to take the axial load. As soon as the power supply is on, the motor shaft momentarily moves, and loads or unloads the ball of the load cell depending on the direction of the axial force before it loads the bearing of the motor. This indicates the load on the bearing. Thus, the axial force on the motor bearing was experimentally determined for different AC motors under the following conditions.

10.4.1 Motor type MC 75288 H 4A 2H

- The motor, without assembling external fan and cooler, was used to measure the axial force being experienced by the bearing at the drive end (DE). It was found that the rotor is pulled towards the drive end with a force of 50 N.
- The same motor, on assembling with an external radial fan and cooler, indicated the axial force on the bearing as 30 N.
- On shifting the rotor of the motor by 1 mm axially towards the NDE (non-drive end) to cause a deliberate asymmetry, and subsequently to change the axial force acting on the bearing, the axial force experienced by the motor bearing was measured as 35 N.

10.4.2 Motor type AC 13248 H 8D

- The measurement of the axial force at the DE of the motor showed that the axial force of 60 N acts towards the NDE as soon as the motor was switched on.
- Furthermore, no significant change in the axial force was noticed on changing the direction of rotation of the motors by altering the phase sequence.

10.5 Results and discussion

The investigations on different motors indicate that the axial force under 'no-load' conditions at the normal speed vary between 30 to 60 N. The direction of the axial force can be either on the DE or NDE depending on the design and type of asymmetry existing in the motors. It has been noted that the presence of a radial fan, cooler and also the existence of asymmetry in the rotor affect the axial force acting on the bearings. The effect of the operating cycle and the axial force produced by the driven equipment are very significant to the load on the bearings of a motor, which might affect performance of the bearings considerably. However, under 'no-load' conditions the axial force on the bearings is insignificant and its low value has no effect on the behaviour and operating life of the bearings. This also confirms that the rotors of the motor are more or less symmetrical, both radially and axially, and that the motors have been manufactured to reasonable accuracy.

10.6 Conclusions and future studies

Based on this study, the following conclusions are drawn:

- The axial force on the motor bearings under 'no-load' conditions is approximately 60 N. It may be either on the NDE or DE. This insignificant axial force does not affect the bearing life and its performance. This also confirms the 'quality' and the manufacturing accuracy of the AC horizontal motors. The premature failure of the bearings on site under 'load' conditions coupled with driven equipment may be attributed to the various external factors, operating cycle and certainly not by the 'axial force' on the bearings generated by the motors themselves.
- A similar scheme may be developed to measure the axial force on the bearings under 'load' conditions to assess the load experienced by various motor bearings, and also to analyse the causes of the premature failure of the bearings.

10.7 Bibliography

1 Binns, K.J. and Dye, M., 'Identification of Principal Factors Causing Unbalanced Magnetic Pull in Cage Induction Motors', *Proc. IEE*, **120**(3), 349–354, March 1973.
2 Ecbert, J. and Gahleitner, A., 'Experimental Method for Determining the UMP in AC Induction Motors', *ERA Trans.* – 2792, July 1970.
3 Vonkachne, P., *UMP in Rotating Electric Machine – Survey of Published Work*, ERA Report z/T 142, 1963.
4 Rai, G.B., *Air Gap Eccentricity in Induction Motors*, ERA 74-1188, July 1974.
5 Rosenbery, E., 'Magnetic Pull in Electric Machines', *AIEE Trans.*, pp. 1069–1113, 1918.
6 Prashad, H. and Shastry, K.S., 'Performance Evaluation of A.C. Motors through Experimental Determination of Axial Force on Bearings', *Condition Monitoring Seminar*, Ahmedabad (India), pp. MF 15–20, 1986.

11
An analysis of the progressive increase in vibration of a large synchronous electric motor

11.1 A general review

In this case study, the various causes of possible vibrations are discussed, in particular, the related causes of a progressive increase in vibrations of a large synchronous electric motor. In addition, the effects of various attempts of balancing on vibration response characteristics of the motor (after eliminating the possible causes of vibrations) are studied and the reasons for the unusual vibration behaviour of the motor and instability of the rotor even after the balancing trials are investigated. On close examination of the rotor of the motor, it is confirmed practically that such behaviour occurs due to thermal effects causing sliding and breakage/loosening of the Bakelite linings used in tightening the salient poles of the rotor.

The particular vibration pattern/characteristics analysed in this case study can be used to diagnose the causes responsible for the vibration problems of any large synchronous motor.

11.2 Introduction

Vibration measurements in large motors are essential for a variety of reasons. The pattern of vibrations in motors is different from other rotating equipment because of the presence of a synchronously rotating magnetic field. Vibrations may be the cause or only the symptom of a problem. In fact, the pattern of vibrations is often a good indication of the condition of a motor. Excessive vibrations or a change in the vibration pattern can be indicative of a serious problem. Vibration levels of a machine should be limited. Excessive vibration levels of a machine indicate various problems associated with design, manufacture, final assembly and installation. In general, in a rotating machine, the change in vibration pattern with time, if detected, calls for diagnosis and thorough investigations.

A vibration process is the conversion of one form of energy into another and this conversion is repeated at regular intervals. The vibrations of an electric motor are both complex and non-harmonic. Therefore, to analyse a vibration problem of a large electric motor, the motor vibration test is conducted from no-load to full-load condition, and from ambient temperature to heated condition, of the motor. The background of the problem of the synchronous motor investigated is as follows. The motor vibration levels were found to be as high as 75 μm (peak to peak) at the rated speed of 1500 rpm. However, during initial commissioning, the vibration levels were 30 μm. The vibration levels were increased to 75 μm in about 500 h of total intermittent operation, which is beyond the acceptable limit. The problem was analysed and the vibration analysis was carried out to investigate the problem and to take corrective measures.

In this case study, the vibration data of the synchronous motor under no-load and partial/full-load conditions in different attempts of balancing are analysed. Routine checks indicated that the motor installation, assembly, coupling, alignment, etc. were in order. However, vibration levels of the motor were found to increase with time. Also, the motor was found to behave in a similar manner under the influence of balance weights in different locations during single and two plane balancing and there was no change in phase angle. The vibration levels increased from as low as 7 μm to as high as 65 μm in a span of 1 to 10 h for both drive end (DE) and non-drive end (NDE) bearings. However, the temperature rise of the bearings remained within the permissible limits (70 °C maximum). The detailed analysis that went into the identification of the cause of the time-bound increase in the vibration levels of the motor is outlined.

11.3 Possible sources of vibration

Among a number of vibration sources in an electric motor, the most common are:

- mechanical imbalance of the rotor;
- magnetic imbalance of the rotor;
- the critical speed of the rotor approaching the rotational speed;
- excessive bearing play and bad bearing mounting and/or lubrication;
- deflection of the motor shaft;
- sticking of the oil film – the phenomenon of squeezing of the lubricating oil film, causing tearing of the Babbitt liner at the high 'points' contact under the long period of standstill condition of the heavy rotor of a large motor.

In an electric motor, the mechanical excitation forces originate from imperfections such as mechanical rotor imbalance, loose rotating parts, bent shafts, misalignment, bad bearings and asymmetry in shaft stiffness. The excitation forces due to these causes become predominant under high temperature and load. Excitation forces due to misalignment with the driven equipment and imperfect coupling can also cause vibrations. In general, a combination of mechanical, magnetic and various external effects cause excitation and increase in vibration levels if proper care has not been taken during installation and commissioning.

Normal clearance on sleeve bearings gives better lubrication by oil ring and oil pressure lubrication. However, excessive clearance causes oil leakage and vibration. Excessive bearing clearance may originate through wear and tear of the bearing liner during normal operation and also by wrong selection of the liner when the type and viscosity of oil are not matched with the liner. Sometimes, unmatched oil grade and bearing clearance are the sources of excessive vibrations and temperature rise.[1] Poor bearing mounting may be another source of vibrations, which shortens the life of the bearing liner.

Shaft bend beyond a certain limit is not permissible; otherwise, dynamic balancing may not be possible and excessive vibrations may persist. A motor shaft with welded spider to carry the laminations can be another reason for shaft bending. Welding stresses, unequal heat transfer from the laminated rotor to the spider or an asymmetrical spider construction may cause the shaft to bend. In this case, the vibration pattern is often a function of the rotor temperature.[2,3]

It is mainly motors with heavy rotors that, if operated periodically with long gaps, are subject to stickiness of the oil film, which may damage the bearing liner and activate excessive vibrations. Stickiness of the oil film occurs when, during long periods of motor standstill, the rotor shaft squeezes out the lubricating oil film. This results in a tearing of the liner at the high load 'points'. It is therefore better to use a jacking pump or a barring motor for the rotors of weights exceeding 4t.[2] Also, the number of starts and stops of the large motors with heavy rotors affects the shear strength and fatigue life of the bearing liner, and may increase the non-bonding area of the liner with the bearing shell. Damping characteristics of the bearing will be inadequate if the bearing liner has excessive porosity. Also, the damping will be insufficient if the bearing is lightly loaded. This may start the particular vibration characteristics of the hydrodynamic journal bearings of an electric motor.

11.4 Diagnosis of causes of vibrations

Various authors have given details of the nature and causes of vibrations.[4] However, there could be entirely different types of causes of vibration for a particular frequency. In general, pure imbalance vibrations would always occur at frequency 1 × rpm. A difference in the stiffness of bearing supports can excite higher multiples. Thermally unstable forgings, wrong assemblies, etc. can be the cause of vibrations due to rotor imbalance. Misalignment and coupling deficiencies have dominant frequencies of 1 × rpm, 2 × rpm and

3 × rpm. Axial vibrations can develop due to a non-uniform magnetic pull during rotation, causing imbalance of the magnetic centre. An unbalanced magnetic pull appears at 2 × supply frequency, irrespective of the number of poles of the motor.[5]

The vibration pattern due to imperfections in the foundation and levelling would generally be inconsistent and unpredictable. It would change with operating parameters, and even direction characteristics cannot be predicted. For defects associated with the bearings, vibrations in velocity mode give a better understanding of the bearing condition.[6]

Another cause of vibration may be the fact that the rotating parts of the machine have a natural frequency equal to the operating speed, which may activate resonance. The vibrations may also be aggravated if the natural frequency of the structural static parts of the machine is much less and coincides with the rotating speed frequency.

11.5 System design, bearing assembly and characteristic features of the synchronous motor under investigation

The bearings used in the synchronous motor under investigation were of the pedestal type, size 254 mm × 381 mm with flood lubrication (external lubrication pressure (LP) system). The grade of oil used was IOC Servo Prime 57. The oil in the pedestal reservoir was 129.6 l, and the flow rate of oil in each bearing was 45 l min^{-1}.

The synchronous motor was used to drive a centrifugal compressor, and was of a revolving field type having solid salient poles, with a closed ventilation circuit. It was air-cooled by water, and the coolers were mounted on two sides of the motor frame. The rotor was supported on two flood-lubricated pedestal sleeve bearings. The whole machine was enclosed in a housing and air-sealed.

The shaft, spider and pole body was an integral steel forging. The pole tip was a high-strength steel forging screwed to the pole body by steel screws. The pole tip kept the field coil tightly pressed to the spider to prevent any movement during running. Bakelite linings were used on the field coils to ensure proper tightening and insulation. Figure 11.1 shows the pole assembly arrangement. Fans were provided, one at each end, to circulate an adequate quantity of air to cool the machine.

The bearing liner was made of steel shell in two halves, each half being lined with tin-based Babbitt metal. The bearing pedestal had an oil reservoir at the bottom. Both bearings were insulated and the pedestal at the driving end was earthed to prevent circulating shaft currents.

The bearing had a low-pressure flood lubrication supplied from an external lubrication system. Two brass rings rotating on the shaft provided oil to the

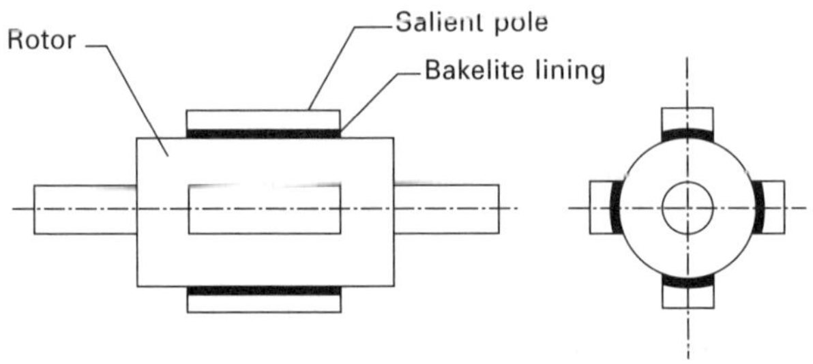

11.1 Original salient pole assembly with Bakelite lining on the rotor of the synchronous motor under investigation.

bearings to prevent them from oil starvation in the event of failure of the external lubrication system. Each bearing was provided with a platinum resistance type temperature detector and a suitable pad for mounting the vibration-measuring probe.

11.6 Investigations and analysis

11.6.1 Vibration characteristics during balancing trials

The motor alignment, bearing clearance and magnetic centre were checked before monitoring the vibration characteristics. In all respects, the motor mounting, installation, its foundation, coupling and lubrication system, etc. were found to be in order.

The vibration analysis has shown predominantly the 1 × rpm component under load and no-load conditions. Attempts were made to balance the motor by single-plane balancing, and then by two-plane balancing using IRD-880, microprocessor, analyser and balancer.

Initially, on adding the trial weight, the motor responded. The change in phase was indicated and vibration levels reduced. However, gradually, the vibration levels were found to increase. In all the balancing trials, on attaining the vibration levels of 50–60 μm (1 × rpm), the phase angle remained unchanged. Both NDE and DE bearings had shown identical behaviour under different balancing trials after about 9–10 h of operation. However, the pattern and rate of increase of vibration levels with time were different. Table 11.1 indicates the similar behaviour of the motor on addition of trial weights at different locations. Although the trials were made by adjusting the clearances in the bearings, the behaviour and vibration pattern of the motor remained unchanged under different balancing trials. Furthermore, no significant change in the bearing temperatures was noticed by adjusting radial clearance and clearance of spherical seating of the bearings.

Table 11.1 Variation in radial vibration levels (peak-to-peak) and phase angle on housing of DE and NDE bearing with time (1 × rpm component) under different modes of trials during balancing

Duration after start (h)	Without balancing weight		Adding 185 g at location 13–14 in plane I		Adding further 200 g at location 3–4 in plane II	
	DE µm/θ	NDE µm/θ	DE µm/θ	NDE µm/θ	DE µm/θ	NDE µm/θ
	48/16	33/17	30/9	12/6	35/11	7/11
1	41/17	31/17	6/8	7/1	16/12	2/11
2	39/17	28/18	8/6	12/1	12/12	4/11
3	45/16	30/17	16/20	15/1	17/14	12/15
4	50/16	35/17	18/20	16/1	20/18	20/19
5	42/16	36/17	23/20	19/1	28/19	22/20
6	54/16	36/17	38/20	29/20	37/19	30/20
7	49/16	35/17	45/19	35/20	42/19	38/20
8	52/17	40/18	55/17	40/18	56/19	49/18
9	60/17	48/17	60/17	48/17	62/18	52/17
10	64/17	50/17	65/17	50/17	66/17	52/17
O/A vibration levels	73	57	75	60	75	58

(i) 360 divided in 20 equal locations (θ).
(ii) θ corresponds to location of phase.
(iii) Vibration levels are at full load.
(iv) O/A corresponds to overall vibration level after 10th hour of operation.

11.6.2 Bearing examination – visually, ultrasonically and by dye penetration

After balancing trials and clearance adjustment in the bearings to resolve the problem (as discussed in Section 11.6.1), the bearings were dismantled and examined. No significant defects were noticed on the surfaces of both NDE and DE bearings. However, no non-uniformity on the surfaces was noticed. The ultrasonic examination revealed a non-bonded area on the lower and upper half surfaces of both the bearings. Besides this, the dye-penetration test indicated that cracks on surfaces of the lower half have DE and NDE bearings originating from the lateral surfaces. Also, a number of scattered and concentrated pin holes were detected on the cylindrical load-carrying surfaces of the bearings, the diameters of the holes varying from 0.5 to 1 mm. In general, the surfaces of the bearings were full of pinholes, and were found to be particularly concentrated at a few locations. The existing bearings were then replaced by new bearings after thorough examination by ultrasonic and dye-penetration tests.

11.6.3 Vibration characteristics after incorporation of new bearings

Even after precise adjustment of radial clearance and clearance of spherical seating of the new bearings, as per the plant standard, no change in the motor behaviour was noticed. The vibration levels were found to increase with time, as earlier. However, the rise of temperature of the bearings was less as compared with the earlier data (<65 °C). Analysis of vibration data indicated the 1 × rpm component to be predominant, although 2 × rpm and 1/2 × rpm components were also present. All the frequency components were inconsistent and found to vary with time. The phase angle was not stable. To resolve the problem, balancing was again tried, but this did not yield consistent results. An increase in the overall vibration levels up to the trip limits (75 μm) was recorded in the span of 9–10 h, as earlier. Furthermore, in a hot start (within 30 min after the stop), the motor picked up the vibration levels more or less from the levels at which it was stopped. However, in a cold start, irrespective of any balancing weights, the motor started with low vibration levels (as low as 7 μm) and picked up the vibrations gradually with time.

11.6.4 Further analysis

After attending to the bearing problems and analysis of the vibration characteristics, it was evident that the problem was of rotor instability. This was also corroborated because of the presence of 1 × rpm component predominantly along with the presence of 2 × rpm and 1/2 × rpm components, besides the system being not responsive to the balancing. This was confirmed practically by removing the stator of the motor and examining the salient poles of the rotor. Figure 11.2 indicates the condition of the salient pole assembly along the length and circumference of the rotor. It is evident that the Bakelite lining of the pole assembly is bulging out non-uniformly and

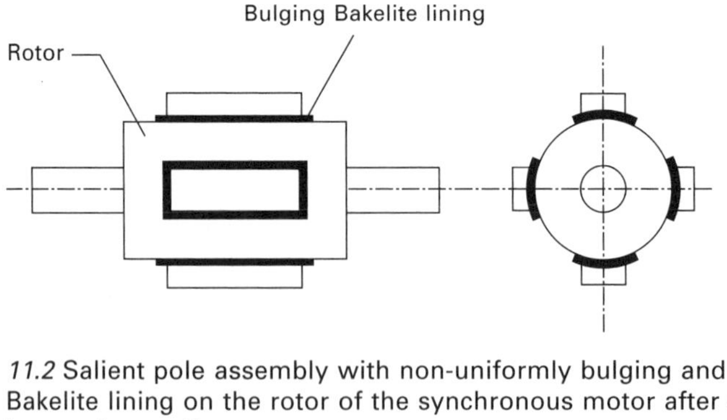

11.2 Salient pole assembly with non-uniformly bulging and sheared Bakelite lining on the rotor of the synchronous motor after a few hours of operation.

heterogeneously in different sections owing to sliding and shearing under thermal effects.

11.7 Results and discussion

11.7.1 Non-bonding of bearing liners

The causes of the larger area of non-bonding detected on the bearing liners may be attributed to the heavy weight (18 t) of the rotor, which, during starting and stopping of the motor, exerted much higher shearing force at the bonded surface. Owing to an increase in the number of starts and stops of the motor, the bonding strength at the bonded surface reduced, and this gradually increased the non-bonded area with time. The higher thickness of the Babbitt liner also added to this and increased the shearing force at the bonded junction compared with the lower thickness of the liner under the same driving force exerted at the instant when the motor was started. Once the non-bonded area had formed, it increased in the subsequent starts and during operation of the motor due to thermal effects. The initiation of the non-bonded area may also be attributed to the long duration of standstill/non-operation of the motor. Under such conditions, the rotor shaft squeezed out the lubricating oil film, which resulted in a tearing of the lining at the high 'points' contact. Starting such a motor without first barring over will cause initiation of permanent damage of the bearing liner and may activate excessive vibration. The formation of the non-bonded area in the motor bearings, attributed to the above reasons, might have damaged the bearing liners in a span of operation after commissioning of the motor, besides the initial quality of the liners.

11.7.2 1 × rpm vibration component

From the data given in Table 11.1, it is evident that the 1 × rpm component of the vibration spectrum is not due to the existing mass imbalance. Had it been, the 1 × rpm component could have been reduced by different trials. However, the effect of the balancing weights proved to be ineffective after a few hours of run. Owing to the balancing weights, initially during 1 to 2 h, the peak-to-peak vibration level had reduced to as low as 2 μm on the NDE bearing, but gradually in the continuous run, the same had increased to 52 μm. Also, there was variation in the phase angle (6 to 20 divisions, i.e. 252°) during operation of both NDE and DE bearings. This peculiar behaviour indicates that the 1 × rpm component is not due to the existing mass imbalance. Furthermore, the response of balancing trials was misleading in the initial operation of the motor (Table 11.1). The 1 × rpm component was predominant throughout the operation and was found to be as high as 66 μm as against 75 μm of overall vibration levels. The variation of 1 × rpm component further reinforces the inference that the initial existing mass imbalance is not the cause of the increase in vibration. However, the variation of rotor stiffness and unbalanced rotor forging under thermal effects in the magnetic field has not been ruled out.

The unusual behaviour of variation of vibration characteristics with time has been proved to be associated with the bearing liner characteristics, when all other checks including misalignment, lubrication, vibration of foundation/pedestal, coupling and its mounting, bearing clearance, etc. were found within limits during investigations. Moreover, on replacing the old liner with a new one, the problem persisted. This led to an investigation on rotor instability.

11.7.3 Investigation on rotor instability

The increase in 1 × rpm component with time was attributed to the increase in temperature of the rotor with duration of operation. As the temperature increased, the poles of the rotor tended to expand but their expansion was restricted by the Bakelite lining used for tightening the pole strips, because of its lower coefficient of thermal expansion. Furthermore, owing to improper tightness and inadequate stiffness, the Bakelite lining slid out from the pole strips and gradually bulged out non-uniformly (Fig. 11.2). Also, because of centrifugal forces and thermal effects, the Bakelite lining was sheared off and found to be crushed at some locations, which initiated cracks on the liners. Thus, thermal shocks due to intermittent starts and stops of the motor gave rise to gradual looseness of the Bakelite lining, which originated 2 × rpm along with 1 × rpm component of vibration at times during the running span of the motor and led to instability of the rotor. Also, owing to the gradual sliding of the Bakelite lining to the rotor periphery, the progressive increase in 1 × rpm component, indicating instability of imbalance and phase shift, was obvious (Table 11.1).

The change in vibration level of 1 × rpm component from 2 to 66 μm with time confirmed the above phenomenon. Furthermore, the initial change in vibration levels after the start of the motor indicated the effect of inadequate temperature rise. Under these conditions, the looseness of Bakelite lining temporarily diminished; thereby, vibrations were slightly reduced owing to an increase in the contact area of the lining with the salient poles. The decrease in vibration levels from 35 to 16 and 12 to 7 μm, explains this phenomenon. However, with an increase in the operating period, irrespective of balancing weights, the vibration levels went on increasing (6 to 65 μm) as shown in Table 11.1.

At times, sudden decreases and then increases in vibration levels (5–8 μm) have been noticed. This behaviour of the motor indicated the effect of different levels of sliding of the Bakelite linings in different planes of the rotor along the rotor length. The sudden decrease in vibration levels from 54 to 41 μm and then increase to 52 μm confirmed the above behaviour (Table 11.1). In hot-start conditions, the rotor retained its profile and so no appreciable change in vibration pattern was detected as compared with the pattern at which the motor was stopped.

During cold start, every time when the motor was started, it picked up the vibrations at different levels. Besides this, owing to the gradual change in the lining position on the rotor, the analogue and digital values of vibration levels on IRD-880 were never stable at any instant. This also led to initiation of the turbulence in oil flow and change in oil film thickness of the bearing due to variation in thermal and mass imbalance of the rotor. Also, hydrodynamic bearings of the motor were found to be lightly loaded due to higher length to diameter (*L*/*D*) ratio (1.5), resulting in inadequate damping of the rotor.

11.8 Conclusions and recommendations

Investigation and analysis of the progressive increase in the vibration levels led to an inference, that was found to be in line with the findings after close examination of the rotor of the motor. The rotor was found to be having non-uniform slippage, accompanied by sliding and breakage of Bakelite lining used in tightening the salient poles of the motor, due to thermal effects, causing imbalance of the rotor.

From the various aspects of the above investigations, the conclusions and recommendations are:[7]

- If the 1 × rpm component during vibration analysis appears and, on different balancing trials, if it still persists and the vibration level increases with duration of operation, then the problem with the motor is attributed to the rotor instability due to thermal effects, causing time-bound imbalance. However, before attributing to this cause, the usual checks pertaining to misalignment, coupling, lubrication, foundation and mounting are to be made to ensure that no such problems exist with the motor.
- The problem with the rotor of a salient pole motor due to sliding and shearing of Bakelite linings on the pole strip exists because of thermal effects, when a progressive increase in 1 × rpm and 2 × rpm components of vibrations are detected and a change in the progressive phase shift is observed.
- If the Bakelite linings connecting the pole strip are loose, bearing vibration level decreases initially and then gradually increases with time.
- In the case of problems associated with Bakelite lining, initial vibration levels of the balanced motor are generally normal. However, with duration of operation, vibration levels increase to 75 μm (after 1–10 h of operation). This is independent of the location of the balancing weights during different balancing trials. However, the rate of increase of vibration levels and

intermittent vibration characteristics during balancing trials differ from each other.

- If the problem of rotor instability due to Bakelite lining of salient poles exists, then every time a motor is started under cold conditions, it starts with different minimum vibration levels. In contrast, in hot start, the motor picks up vibrations from more or less the same levels at which the motor was stopped.
- During shut-down/overhaul, the bearing liners of large electric motor should be examined ultrasonically, and the non-bonded area and cracks in the liners should be monitored.
- Poles of the motors should be adequately tightened. The quality of the Bakelite lining should be assured. The motor should be tested for at least 10 h to ascertain the stability of the rotor. The investigations and analysis given in this case study can be applied to diagnose the problems associated with a large salient pole synchronous motor, and to tackle problems associated with the rotor instability.

11.9 References

1 Prashad, H., 'The Effects of Viscosity and Clearance on the Performance of Hydrodynamic Journal Bearings', *STLE Trans.*, **31**(1), 113–119, 1988.

2 Nevelsteen, J., 'Vibration, Causes and Effects on Large Electric Motors', *25th Annual Petroleum and Chemical Industries Conference*, New York, pp. 214–217, 1978.

3 Maxwell, J.H., 'Diagnosing Induction Motor Vibration', *Hydrocarbon Proc.*, **60**, 117–120, January 1981.

4 Gupta, S.K., 'Vibration in Induction Motors – Concepts, Nature, Analysis and Control', *BHEL J.*, **5**(1) 1–8, 1981.

5 'Criteria for Assessing Mechanical Vibrations of Machines', VDI 2056, Oct. 1964.

6 Baxter, R.L. and Bernhard, D.L., 'Vibration and Indicating Tool', *Mech. Eng.*, 60–65, March 1968.

7 Prashad, H. and Rai, L.N., 'Analysis of Time Bound Increase in Vibrations of Large Synchronous Electric Motor – An Investigation', *BHEL J.*, **13**, 54–61, 1992.

12

A study of the causes of failure of rolling-element bearings in alternators

12.1 A general review

This diagnosis deals with the causes of failure of rolling-element bearings used in alternators and establishes the reasons as to why the bearings used in a particular design of the alternators failed prematurely. The voltage across the bearings leading to the passage of electric current and the development of magnetic flux density on the bearing elements are experimentally determined, as these cause the premature failure of rolling-element bearings of the alternators. The development of stray voltage and excessive magnetic flux density are established in the particular design of the alternators where the bearings are located under the stator field coil and AC coils, as against those designs where the bearings are housed at a suitable location so as to avoid the effect of excited magnetic flux by the field and AC coils.

The findings reported in this diagnosis give overall guidelines to designers to avoid premature failure of bearings.

12.2 Introduction

The present study was undertaken to diagnose the failure problem, where the identical bearings used in 'A' type alternators were reported as having a high rate of failure as compared with those used in 'B'-type alternators. The cause of failure is diagnosed by the magnetic flux density measurement on the surfaces of the bearings and by measurement of voltage between the bearing and shaft, and shaft ends. Also, the cause of the high residual flux density on the bearing surfaces is analysed for the alternators of 'A' type where the high failure rate of the bearings is reported/established. Besides this, deterioration of used grease and discolouring of the track surfaces establish current flow through the damaged bearings of the alternators.

12.3 Design features of the alternators

The alternator consisted of two sets of winding, namely, AC winding and field winding, both accommodated in the stator. The AC windings were distributed in the small slots and field windings were concentrated into two slots. Each field coil spanned half the total number of stator slots. AC coils were connected in star and field coils were connected in series. The rotor, consisting of stacked stampings, resembled a cogged wheel, having eight sets of teeth and slots, uniformly distributed on the rotor surface skewing the rotor axis.

The stator, which is completely embraced by the field coils, retained a residual magnetism if excited by a battery once. The flux produced by the field coils finds its path through the rotor. When the rotor was rotated, the passage of the rotor teeth and slots alternately under the field offered a varying reluctance path for the flux produced by the field coils. This flux, which varied periodically, linked with the AC coils and induces an alternating voltage in the AC coil. The frequency of the induced voltage depended on the speed of the rotor. The magnitude depended on the speed of the rotor and level of excitation. The field is strengthened by a positive feedback system in the regulator to attain the desired output voltage.

Alternators were used for developing a constant voltage of 110 V, 4.5 kW at the rated load current of 37.5 A from minimum speed to maximum speed. The alternators were used to charge a battery and to operate lighting. The outstanding technical features of the alternators were as follows:

- Output voltage: 110 V ± 5%
- Rated and maximum current: 37.5 and 43 A
- Cut in, minimum and maximum speed: 350 rpm, 550 rpm and 2650 rpm

12.3.1 Design modification of alternators

Earlier types of alternators were designed with an output voltage of 24 V with a 4.5 kW rating. In the recent past, alternator design was modified and output voltage increased to 110 V with the same kW rating, besides the other minor modifications. However, the bearings in alternators of different makes, both in the earlier and in the later modified design, are the same, i.e. NU 311 on the drive end and 6309 on the non-drive end.

12.4 The nature of bearing failure

The failure of driving end bearing type NU 311 was reported in the modified alternator design of type 'A' having 110 V and 4.5 kW rating while the bearing of non-driving end was intact. The nature of failure was the formation of corrugations and ridges on the raceways as well as corrosion of the raceways and rolling-elements of the bearings, irrespective of the make of the bearing used. The bearing failure was not reported in the earlier design of alternators of type 'B', i.e. 24 V and 4.5 kW rating.

12.5 Data collection and investigations

12.5.1 Failed bearings and lubricant

Failed bearings of type NU 311 of the alternators from various sites were collected. Almost all the failed bearings were from alternator type 'A'. Figure 12.1 to 12.4 show these failed bearings. Samples of the fresh grease and used

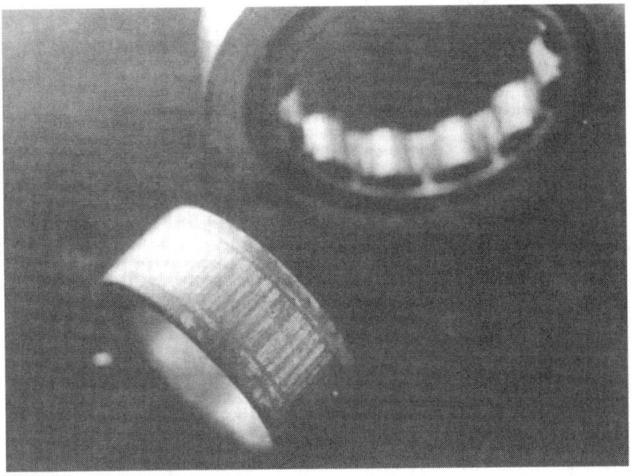

12.1 Corrugation pattern on damaged bearing of alternator type 'A'.

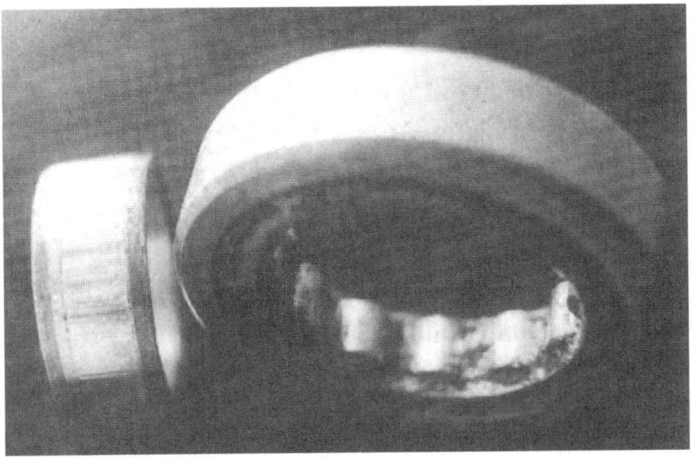

12.2 Inner race and rolling elements of damaged bearing of alternator type 'A'.

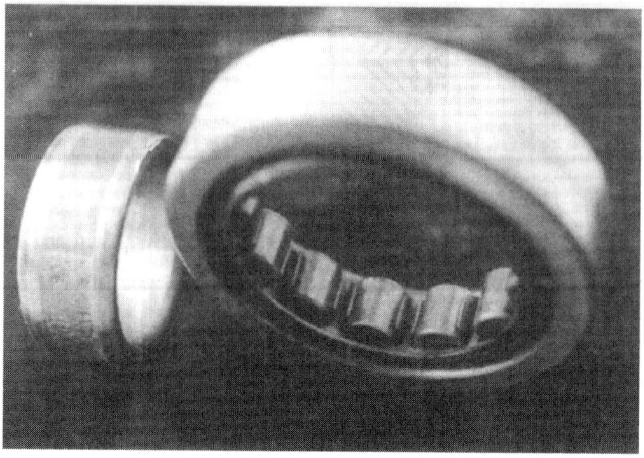

12.3 NU 331 failed bearing of alternator type 'A'.

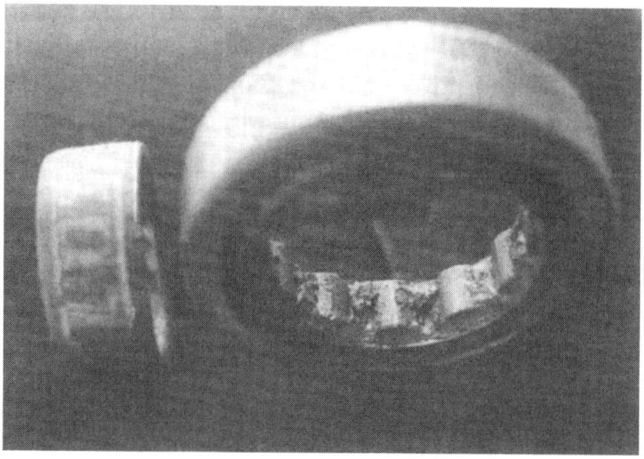

12.4 Corroded inner race and rolling elements of alternator type 'A'.

grease from the rolling-elements of the failed bearings were collected for analysis.

12.5.2 Bearing location in alternator design

The design configuration with respect to bearing locations in the alternator types 'A' and 'B' was studied. Figures 12.5 and 12.6 show these bearing locations. It was found that the NU 311 bearing on the drive end of alternator type 'A' design was mounted in the housing provided in the projection of the end cover. However, the projection was made towards the inside of the alternator, thereby locating the bearing almost under the influence of the

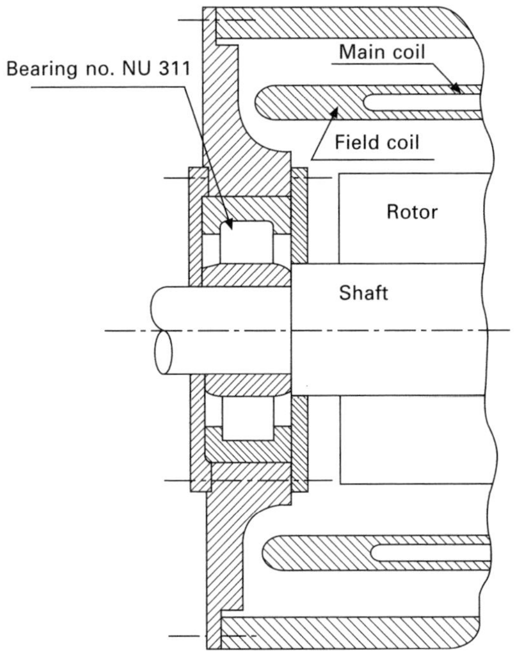

12.5 Design configuration showing location of bearing, field and main coils of alternator type 'A'.

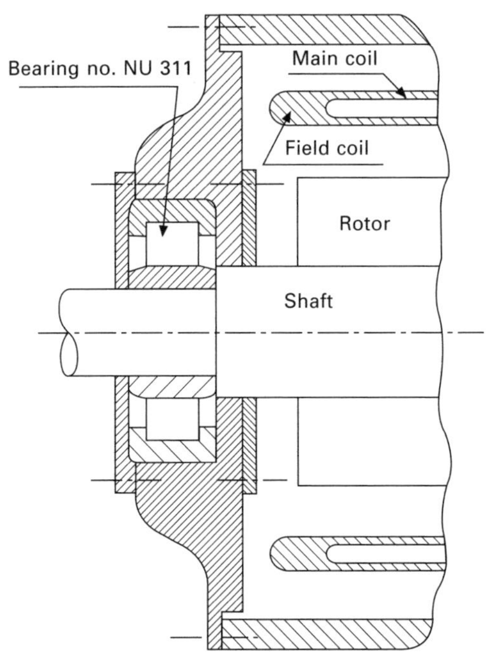

12.6 Design configuration showing location of bearing, field and main coils of alternator type 'B'.

field coil, as shown in Fig. 12.5. The drive end bearing of alternator type 'B' design was mounted in a similar projected housing on the end cover, but the projection was outside the alternator and, hence, the bearing was outside the purview of the field coil, as shown in Fig. 12.6.

12.5.3 Investigations

Measurement of the shaft voltages

The investigations were carried out pertaining to the shaft voltage and stray voltage on the alternators of type 'A' and 'B' and on the high-capacity alternators available at works. Table 12.1 shows the measured data under different load conditions of the alternators.

Measurement of flux density

The magnetic flux density was measured using a Hall probe and Gauss meter on the track surface of the inner race, outer race and rolling-elements of the failed bearings. No magnetic flux density was detected on the surface of new bearings irrespective of the make.

Inspection

Dimensional accuracy of new bearings was checked and found to be in line with the specifications. Also, dimensional accuracy and metallurgical examination of the components of the failed bearings indicate that the results were in line with the specified norms.

Table 12.1 Voltage measurement on different alternators

Alternator at work under full load	
(a) 160 kVA alternator	
Voltage between shaft and body	0.30 V
(b) 500 kVA alternator	
Voltage between shaft and body	0.21 V
Alternator type 'A'	
Stray voltage (no load)	2–6 V
Voltage between shaft and body (no load)	0.9–1.6 V (AC)
Stray voltage (30 A load)	6–20 V (AC)
Voltage between shaft and body (30 A load)	2–6 V
Alternator type 'B'	
Stray voltage (no load)	Nil
Voltage between shaft and body (no load)	0.14–0.21 V
Stray voltage (30 A load)	Nil
Voltage between shaft and body (30 A load)	0.14–0.30 V

12.6 Causes of shaft voltage and flow of current through bearings

Shaft voltage exists in electrical machines as a result of asymmetry of faults, winding faults, unbalanced supplies, electrostatic effects, air-gap fields, magnetized shaft or other machine members, asymmetries of the magnetized fields, etc.[1] The causes of shaft voltage can be grouped under four categories, namely:

- external causes;
- magnetic flux in the shaft;
- homopolar magnetic flux; and
- ring magnetic flux.

Friction between belt and pulley can set up the electrostatic voltage between shaft and bearings, which acts as an external cause. An accidental grounding of a part of the rotor winding to the rotor core can lead to stray currents through the shaft and bearings and can result in the permanent magnetization of the shaft. Also, the shaft voltage and current could be generated when the machine is rotated. Besides this, homopolar flux can result from an air gap or rotor eccentricity, and this can generate voltage.[2]

The most important cause of bearing current was the linkage of alternating flux with the shaft. The flux flowed perpendicularly to the axis of the shaft and pulsated in the stator and rotor cores. It was caused by asymmetries in the magnetic circuit of the machine, such as:

- uneven air gaps and rotor eccentricity;
- split stator and rotor core;
- segmented punching;
- axial holes through the cores for ventilation or
- clamping purposes;
- keyways for maintaining the core stacking;
- segments of different permeability.

All the causes listed above developed the magnetic flux, which closed in the circumference over the yoke and induced voltage on the shaft as the machine rotated. This resulted in a localized current at each bearing rather than a potential difference between shaft ends. A current path, however, along shaft, bearings and frame resulted in a potential between shaft ends.[1,3]

At a certain threshold voltage depending on the resistivity of the lubricant and operating conditions, current flowed through the bearing.[4] Thus, the flow of circular current in the inner race leaked through the rolling-elements to the outer race by following the path of least resistance and established the field strength, leading to the development of magnetic flux on the track surface of races and rolling-elements.[5]

12.7 Results and discussion

12.7.1 Failed bearings analysis

The corrugations and ridges found on the track surface of races of all the failed bearings besides the corrosion on the surfaces, as shown in Fig. 12.1 to 12.4, indicate exposure of bearings to electric current.[1,3,6] The corroded surface of the bearings occurred because of the decomposition of the grease, and formation of corrugations on the bearing surface by low-temperature tempering and Hertzian pressure on the raceways,[6,7] which finally resulted in a reduction of bearing fatigue life. The phenomenon of the formation of corrugation is explained as follows.

At each revolution of the shaft, part of the circumference of the inner race passed through the zone of maximum radial force, and Hertzian pressure between rolling-elements and raceways at the line of contact, led to maximum shear stress. Maximum shear stress is taken as the criterion for yielding and this occurs in the subsurface at a depth approximately equal to half the radius of contact surface. It is generally at this point that the failure of material, if occurring, will initiate.[1,3]

The above process of deformation leads to the formation of a corrugation pattern on the bearing surfaces and is accelerated by the passage of current, corrosion by decomposition of the lubricant, lubricant characteristics and quality of the bearing. Based on this, corrugation first forms on the track surface of the races and then on the rolling elements. Misalignment of the alternator due to over-tensioning of the belts and imbalance of the shaft, if any, may accelerate formation of corrugation and damage on the bearing surfaces.[1,7]

12.7.2 Magnetic flux density

The magnetic flux density on the track surface of races was measured to be as high as 40G at certain locations. In general, the magnetic flux density varied between 10 and 40G.

The presence of magnetic flux density along with the corrugation pattern and corrosion on bearing surfaces indicates the damage due to electric current.[5,8] Merely the presence of a corrugation pattern without significant flux density distribution indicates plastic deformation after the fatigue damage of the surface by mechanical loading, accompanied by the flexibility of the supporting structure. This deformation is influenced by the frequency of rotation of rolling-elements and inner race.

Very frequently, the mode of failure of the electrical contact is an adhesive wear or the transfer of material from one surface to another. Generally, a pit is observed in one surface and an amount of material is observed to have transferred to the other due to the contact. Under the influence of an electrical current, an interface in a tribological system tends to provide more energy to promote adhesive wear and transfer.[5]

Effects of residual magnetic flux on bearings

The residual magnetic flux thus detected on the surface of the bearings can lead to premature failure through mechanical forces or the heat generated. The flux passing through the bearing produces forces of attraction, hysteresis loss and eddy current. These parasitic energy losses are supplied from the mechanical shaft power, and, therefore, increase the torque required for turning the shaft. The higher temperature as against the temperature of bearings operating under rolling-friction without the influence of magnetic flux density confirms the increase in parasitic energy losses.[9]

The relative velocity of a rolling-element with respect to inner race is zero at the point of asperity contact, but increases linearly with distance from the point of contact. Thus, the races will have an area of high flux density on the surface at the line contacts, which induces voltage circulating eddy currents in the face of the race from the area of high to low flux density.

The hard steel bearings have relatively large magnetic hysteresis losses. The loss can be due to a variation in magnitude of flux. During rotation of a bearing, load is distributed among a limited number of rolling-elements. This may further lead to an instant change in fluxes and, hence, cyclic change in attractive forces on bearing elements. This may cause premature failure of the bearings due to non-uniform wear of the bearing surfaces.

Origin of magnetic flux density and voltage

From Fig. 12.5, it is evident that the bearing NU 311 is located under the field winding in the design of alternator type 'A' as against that of alternator type 'B' as shown in Fig. 12.6. This makes the NU 311 bearing of alternator type 'A' permanently magnetized and a magnetic flux density develops on the bearing elements.[8] Since the NU 311 bearing in alternator type 'B' is located away from the influence of field coils, the magnetic flux density is not developed on the bearing elements. The experimentally determined magnetic flux density of the order to 10 to 40G on the bearing elements of alternator type 'A', as against the negligible flux density of alternator type 'B', confirms this phenomenon.

When the outer race of the bearing is magnetized and inner race and rolling-elements rotate inside the outer race, the voltage is generated by electromagnetic principles and the flow of current starts through the inner race and rolling-elements depending on the impedance of the oil film thickness and the threshold voltage phenomenon. It is confirmed by the stray voltage of 2–6 V and 6–20 V measured on alternator type 'A' under 'no-load' and 'load', conditions, respectively. In contrast, in alternator type 'B', no stray voltage was detected (Table 12.1).

Besides the stray voltage phenomenon, the voltage between shaft and body under 'no-load' and 'load' was measured as 0.9–1.6 V and 2–6 V, respectively, in alternator type 'A' as against low voltage in alternator type 'B'. This shows asymmetry of the magnetized field, rotor eccentricity, etc. of alternator type 'A' apart from the other causes given in references 1 and 3. However, in addition to the other manufacturing errors, the major contribution to the higher voltage between the shaft and the body is attributed to the bearing location, which is under the influence of the field coil and is able to magnetize the bearing and damage it in due course.[8] The influence is more significant in the 110 V design of alternator type 'A' than in the 24 V designs. Furthermore, it has been established that the shaft voltage is a function of the kilowatt rating of the machine.[2]

12.7.3 Effect of bearing current on lubricant

The zinc additive, i.e. zinc dithiophosphate or zinc dialkyldithiophosphate (ZDTP) used as a multifunctional additive in the grease, under rolling friction protects the rubbing metal surfaces and contributes to friction and wear reduction, and depends, partly, on the amount of additive on these surfaces. Physisorption and chemisorption processes precede the chemical reactions with metals; therefore, it is probable that the load-carrying capacity is related to these processes. Decomposition of ZDTP in the lithium base grease under the influence of electrical fields leads to the formation of lithium zinc silicate ($Li_{3.6}Zn_{0.2}SiO_2$) in the presence of a relatively high percentage of free lithium and silica impurity in the grease at high temperature in the asperity contacts along with the formation of gamma lithium iron oxide (γ-$LiFeO_2$). Besides this, the original structure of lithium stearate changes to lithium palmitate. These changes are not detected under rolling friction. The used grease taken from the failed bearing has shown changes similar to that shown in reference 6.

12.7.4 Process of bearing failure under the influence of leakage current

When the current leaks through the roller bearing in which low-resistivity ($10^5\,\Omega\,m$) grease has been used, 'silent' discharge passes through the bearing elements. This creates a magnetic flux density distribution on the bearing surfaces. Also, when a bearing is located and operates under the influence of a magnetic field, voltage is generated and current flows through the bearing, depending on bearing impedance and the threshold voltage phenomenon. This leads in the initial stages to the electrochemical decomposition of the grease and corrosion,[6,7] and then gradual formation of flutings and corrugation on the surfaces.[1,3,7] Subsequently, wear increases and the bearing fails.

When high resistivity ($10^9\,\Omega$ m) grease is used, the charges accumulate on the bearing surfaces due to polarization till they reach the critical value at which a feeble current is conducted through the bearing,[4] which is not able to produce any significant flux density distribution on the surfaces. However, mass transfer at the elevated local temperature accompanies this on asperities of the friction surfaces by sudden discharge leading to surface pitting by arc welding effect and failure of a bearing.[5,7]

12.8 Conclusions and recommendations

From the findings of the analysis given in this chapter, the following conclusions are drawn and recommendations made:[10]

- Under the influence of field coils, a roller bearing develops magnetic flux. During operation of a bearing under the influence of magnetic field, voltage is generated. Also, stray voltage on the bearing is developed.
- The current passes through a bearing depending on bearing impedance and threshold voltage phenomenon.
- The varying magnetic flux density of 10 to 40G is developed on the track surface of the races of bearings depending on the flow of current.
- In the case of bearings using grease of low resistivity, failure occurs under the silent electrical discharge due to chemical decomposition, formation of flutings and corrugation on bearing surfaces. With grease of high resistivity, failure occurs through an accumulation of charge and its subsequent breakdown.
- To avoid the failure of a bearing under the influence of electric current, bearings should not be located in an alternator under the influence of field coils.
- In existing machines, bearings should be insulated/shielded so as to avoid the flow of current through them.

By the detection of flux density on the bearing surfaces along with the corrugated pattern on the track surfaces and also by analysis of deterioration of greases used in the bearings[5,7,8] and measurement of stray and shaft voltage, the failure of bearings by the passage of electric current can be established.

12.9 References

1 Prashad, H., 'Investigations of Corrugated Pattern on the Surface of Roller Bearings Operated under the Influence of Electrical Fields', *Lub. Eng.*, **44**(8), 710–718, 1988.

2 Bradford, M., *Prediction of Bearing Wear Due to Shaft Voltage in Electrical Machines*, ERA Technology Limited, 1984.

3 Prashad, H., 'Theoretical and Experimental Investigation on the Pitch and Width of Corrugations on the Surface of Ball Bearings', *Wear*, **143**, 1–14, 1991.

4 Prashad, H., 'Effects of Operating Parameters on the Threshold Voltages and Impedance Response of Non-insulated Rolling-element Bearings Under the Action of Electric Current', *Wear*, **117**, 223–240, 1987.

5 Prashad, H., 'The Effect of Current Leakage on Electro-adhesion Forces in Rolling-friction and Magnetic Flux Density Distribution on the Surface of Rolling-element Bearings', *ASME, J. Tribol.*, **110**, 448–455, 1988.

6 Prashad, H., 'Diagnosis of Deterioration of Lithium Greases used in Rolling-element Bearings by X-ray Diffractrometry', *Tribol. Trans.*, **32**(2), 205–214, 1989.

7 Prashad, H., 'Investigation of Damaged Rolling-element Bearings and Deterioration of Lubricants under the Influence of Electric Current', *Wear*, **176**, 151–161, 1994.

8 Prashad, H., 'Magnetic Flux Density Distribution on the Track Surface of Rolling-element Bearings – An Experimental and Theoretical Investigation' *Tribol. Tran.*, **39**(2), 386–391, April 1996.

9 Tasker, J.L. and Graham, R.S., 'Effects of Magnetic Flux on Rolling-element Bearings', *IEE Conference*, 17–19, Sept. Publication 254, pp. 152–156, 1985.

10 Prashad, H., 'Diagnosis of Failure of Rolling-element Bearings of Alternators – A Study', *Wear*, **198**, 46–51, 1996.

13
The diagnosis of the cause of a bearing problem in a synchronous condenser

13.1 A general review

The investigations reported in this diagnosis concern the effect of the use of deteriorated lubricant on bearing life. The diagnosis of the premature failure of the hydrodynamic journal bearings of a synchronous condenser was carried out. The surface of the failed bearings was examined and the cause of the failure investigated. Also, the theoretical analysis established the number of cycles needed before the level of detected severity of the failure occurred. The number of cycles matched that of the cycles' duration of operation for the premature failure of the bearings. The process leading to the premature failure and role of the lubricant was identified. Also, remedial measures have been successfully implemented to reuse the existing bearings. Measures have been suggested to avoid the repeated failures.

The analysis has the potential to ascertain the shaft voltage once the severity of the physical damage on the liner surface of the bearing and duration of its operation have been established.

13.2 Introduction

The journal bearings of a synchronous condenser were reported damaged approximately after 2 h of run after installation of the re-Babbitted journal bearings. The bearings were lubricated with hydrostatic lubrication facility and were equipped with pressure-fed forced oil lubrication. During examination of the bearings, the surface of the liner was found to be corroded and had craters on the scattered area particularly in the zone of load-carrying oil film/minimum film thickness. Theoretical and experimental investigations were carried out to establish the cause of premature failure and remedial measure suggested.

The theoretical studies bring out the effects of instantaneous leakage of electrical energy between the journal and bearing lubricated with low-resistivity oil having low viscosity due to water contamination, so as to determine the minimum number of cycles/revolutions for the formation of craters on the bearing liner. The volume of craters appeared on the liner surface or shaft voltage across the bearing or the developed model can assess current passing through the bearing. The gradual rise in temperature at the high 'points', because of the passage of current causing softening of the liner at high 'points' contact and reduction in its toughness, is considered for the analysis. The scope of the analysis includes the determination of width of contact between journal and bearing, bearing inductance and energy of inductive circuit and number of cycles for the appearance of craters on the bearing liner.

13.3 Technical details of the synchronous condenser

13.3.1 Design features

The synchronous condensers are used for maintaining the reactive power output within the permissible limits. Design technical features of the synchronous condenser are as follows:

- Output – 30 000 kVA leading with forced excitation and 15 000 kVA lagging with reduced excitation.
- Voltage – 11 000 V with permissible voltage variation of ±825 V.
- Frequency – 50 cycles s^{-1} with permissible variation of ±2.5 cycles s^{-1}.
- Synchronous speed – 750 rpm, over speed 863 rpm with a duration of 5 min.
- Starting voltage – 2000 V.

13.3.2 Bearings and lubrication system

Bearings of 300 mm diameter having $L/d = 1$ lined with the tin base Babbitt, and the bearing shell with spherical seating for ensuring the self-aligning were used in synchronous condensers.

The lubrication was done by oil ring and forced oil circulation. The oil ring was a rather auxiliary method and served especially as a standby lubrication method in case the power source fails. In general, lubrication and cooling of the bearings were done by force oil circulation. The hot oil produced in the bearing flowed to the oil tank. The oil pumps (one for normal use driven by an induction motor and one for emergency use driven by a DC motor) fed the cooled oil from the oil tank to the bearings through the oil cooler.

13.4 Experimental procedure

Within approximately 2 h of starting the synchronous condenser, the vibration levels had increased beyond the permissible limits. On stopping the condenser,

13.1 Crater formation on lower half of damaged bearing.

the bearings were inspected. The lower halves of the exciter and non-exciter end bearings were found to be damaged by the pitting marks around the areas of minimum film thickness. Figure 13.1 shows the lower half of the exciter end damaged bearing, indicating the crater formation on the liner surface. Pittings/craters were observed throughout the bearing length in the zone of minimum film thickness having a diameter of approximately 0.2 mm and depth of craters on the surface approximately 0.12 mm, and were found to be very close to each other in the zone of minimum film thickness as shown in Fig. 13.1. About 1000 craters were measured on the liner surface. The oil grade conforming to BS 489-1974 used for the lubrication was inspected, and functional characteristics analysis, including that of viscosity, viscosity index and flash point, etc. of the fresh oil and used oil, was performed to establish the deterioration of used oil in the bearings. The condition of the strainers/oil filters of the lubrication oil system was examined. Also, ultrasonic examination of both exciter and non-exciter end bearings were carried out to determine the non-bonded area of the Babbitt liner with that of the bearing shells.

13.5 Measurement obtained

From Fig. 13.1, it is evident that pitting/craters formation occurs because of the current passing through the bearing.[1,3] Craters/pitting were found only on the liner surface in the zone of minimum film thickness, and were only a surface phenomenon. The total volume of the craters was determined as approximately $3.77 \times 10^{-12} m^3$ (Fig. 13.1). The shaft voltage, as reported on measurement between the rotating shaft and the earth, was nearly 2.5 V.

13.5.1 Characteristics of fresh and contaminated lubricant

The lubricating oil from the oil tank and in the drainage line of the lubricating oil system was thoroughly inspected and found to be decolourized and having water contamination. The viscosity of the oil was determined and found to be very low. Owing to the contamination of the oil with water as detected, the oil functional properties were completely deteriorated. The oil, because of the water contamination, loses its original resistivity and becomes the current-conducting medium in the bearings. The technical particulars of the fresh and used oils are given in Table 13.1.

13.6 Theoretical

13.6.1 Causes of shaft voltage

Shaft voltages exist in the electrical machines including the synchronous condenser as a result of asymmetry of faults, windings faults, unbalanced supplies, electrostatic effects, air gap fields, magnetized shaft or other machine members, asymmetries of the magnetized fields, etc.[1] They are mainly due to electromagnetic-induction by the unbalanced magneto-motive force (mmf) produced in an asymmetrical magneto-motive circuit. Shaft voltage increases linearly with the rate of change and overall magnitude of a voltage transient, until saturation is reached. This suggests that there is magnetic saturation in some part of the circuit. As the shaft voltage is mainly related to the supply voltage, the overall magnitude of the voltage rather than the rate of change is likely to be the predominant factor. Also, the induced voltage increases with the speed in an approximately linear manner.

Table 13.1 Significant characteristics of lubricants

	Fresh oil (as per BS 489-1974)	Deteriorated oil
Kinematic viscosity (centistoke at 40°C)	64–72	1–2
Viscosity index (min.)	95	–
Flash point, (°C min.)	210	Undetermined
Resistivity (Ωm)	10^{12}	$<10^4$

In general, the magnitude of the shaft voltage depends on the size of the machine, number of coils turns, current, magnetic asymmetry, rotor eccentricity, strength of residual magnetism, number of poles, type of winding, etc. At present, there is no satisfactory relation to obtain the level of shaft voltage and bearing impedance.[1–3] At a certain threshold the shaft voltage, depending on the resistivity of the lubricant, electrical breakdown of the oil film and passage of current through the bearing, is created, which causes crater formation on the liner surface of the bearing by the arcing effect.[4]

Flow of current depends on the shaft voltage and bearing impedance. The impedance is a function of oil characteristics, resistivity, oil film thickness, eccentricity and clearance ratio, length/diameter ratio, capacitance and inductance of the bearing circuit. Capacitance is a function of bearing design and operating parameters whereas inductance is a function of bearing length and clearance/diameter ratio of a bearing.[5,6] The current flows within the shaft and bearings through oil films. The voltage over a bearing may be very low but local current of a very high order might flow depending on the bearing impedance. Bearing impedance is sensitive to temperature, load and speed but remains constant over a wide frequency range of applied voltage.[5,6]

13.6.2 Width of contact between the journal and bearing

Under the influence of load, the width of contact around the circumference between the journal and a hydrodynamic bearing is determined as:[5]

$$W_b = 2.15\,d[p(1 + \psi^{-1})/E]^{1/2} \qquad [13.1]$$

13.6.3 Bearing inductance and energy of inductive circuit

When the current in the circuit is increasing, the induced electromotive force is in the opposite direction to the current. The energy, i.e. the total work done in opposing the electromotive force (emf) while the current I_b is being established in the inductive circuit, is given as:[6,7]

$$Q = \frac{1}{2} L_j I_b^2 \qquad [13.2]$$

where

$$L_j = 2 \times 10^{-9}\, L \ln\,(1 + C_d/d) \qquad [13.3]$$

For the bearing using low-resistivity lubricant, when the resistance of the circuit is low but not zero, there is a continuous dissipation of electrical energy into heat as the current flows. This process is not reversible, the energy gradually reduces and, before it finally vanishes, the electrical energy reappears because of the shaft voltage and induction phenomenon. Thus, the process is continued and the electrical energy is converted into heat and the crater formation on the liner surface gradually takes place.

13.6.4 Number of cycles/revolutions for appearance of craters on the bearing liner

When a bearing lubricated with a low-resistivity oil is operating under the influence of shaft voltage, the charge is inducing by the self-inductance on the inner peripheral surface of the bearing and an equal and opposite charge is induced on the outer peripheral surface of a journal bearing, which is earthed for the non-insulated bearing. The slow passage of current takes place through the bearing when the high 'points' on the journal and the bearing in the zone of load-carrying oil film come close and when the resistance of the oil film is negligible under the influence of the shaft voltage exceeding the threshold voltage.[4] As soon as the conducting paths are broken when high 'points' become separated by the greater thickness of oil film, arcing results owing to instantaneous leakage of stored energy between journal and liner by vibration effects and to various instabilities of journal bearings.

The minimum energy (Q_n) dissipated in N_p cycles leading to the formation of craters on the liner surface is given by:

$$Q_n = \frac{1}{2} L_j I_b^2 N_p \qquad [13.4]$$

This dissipated energy influences K high 'points' (of radius r and depth H) in the zone of load-carrying oil film and increases temperature to T_n in N_p cycles of operation. Furthermore, each high 'point' of the bearing touches the journal ($\pi d/W_b$) times in each shaft rotation. So, using equation (13.4) and neglecting heat loss by oil flow and other means, it is evident that:

$$\frac{1}{2} N_p I_b^2 L_j = \pi r^2 HK\rho(CT_n(\pi d/W_b)\Delta \qquad [13.5]$$

The ratio of duration of growth to duration of decay of bearing current (t_g/t_d) during the process of growth/decay of energy by shaft rotation plays an important role in determining the net cycles to create craters on the liner. So, considering the ratio of t_g to t_d, the minimum number of cycles for specified dimensions of craters is determined as t_g/t_d times the value of N_p and is given as (using equation 13.5):

$$N = \frac{2\pi d V_v \rho C T_n}{I_b^2 L_j W_b}(\Delta t_g/t_d) \qquad [13.6]$$

In general, formation of craters occurs in the zone of minimum film thickness along the bearing length. To determine the minimum number of revolutions for initiation of craters on the liner, localized high 'point' temperature (T_n) of 120 °C, the softening temperature of the liner at which material toughness drastically reduces, was considered for analysis. Furthermore, t_g/t_d depends on the transient to steady-state value of current and varies between 1 and 25. Since the growth and decay of the current depends on shaft frequency and movement of the shaft in and out of the load-carrying oil film, so t_g/t_d is taken depending on transient to steady-state value of bearing current for the analysis.[6]

13.7 Comparison of theoretical and experimental data

Based on the volume of craters measured on the bearing liner, shaft voltage, bearing resistance and bearing dimensional parameters as shown in Table 13.2, the net number of cycles before the formation of craters on the liner surface, was determined using equation (13.6). The number of cycles, thus determined theoretically (12×10^4) was found to match closely with the experimental site data (9×10^4 cycles) of the operation of the machine before the bearings of synchronous condenser was dismantled and examined for investigations:

- Number of cycles/revolutions before formation of craters on the liner surface of the bearing (experimental) = 9.0×10^4 (approx. 2 h).
- Number of cycles/revolutions before formation of craters on liner surface of the bearing (theoretical analysis) = 12×10^4 cycles (2.67 h).

Table 13.2 Bearing dimensions, operating parameters, investigated and analytical values

d	= 300 mm
L	= 300 mm
n	= 750 rpm
C_d/d	= 0.0015
p	= 2×10^6 N/m^2
H	= 0.12 mm
r	= 0.1 mm
V_v	= 3.77×10^{-12} m^3
K	= > 1000
ρ	= 7.46×10^3 kg/m^3
L_j	= 9×10^{-13} henry (V s/A)
W_b	= 0.118 m
C	= 2.26×10^2 W s/kg °C
E	= 40×10^9 N/m^2
Δ	= 0.1 (0.1 to 0.5)
t_g/t_d	= 1.4
V	= 2.5 V
R_s	= 0.02 Ω
I_b	= V/R_s = 125 A

13.8 Results and discussion

13.8.1 Number of cycles for formation of craters on liner surface

The minimum number of cycles for formation of craters (N), of specified volume on the liner surface depends on bearing dimension, specific load on bearing and liner properties and is inversely proportional to the square of the bearing current and the bearing inductance as shown in equation (13.6). This is also a function of the ratio of duration of growth to duration of decay of bearing current (t_g/t_d) and the ratio of asperity to the net contact surface area of the bearing liner and journal in each shaft revolution (Δ) (Equation 13.6).

For a 300 mm diameter bearing having (L/d = 1) and inductance (L_j) of 9×10^{-13} H operating at 750 rpm with $p = 2 \times 10^6$ N/m^2, the number of cycles to form craters of volume (V_v) 3.77×10^{-12} m^3 takes 12×10^4 cycles, under the influence of shaft voltage of 2.5 V and bearing resistance (R_s) as 0.02 Ω. The current passing through the bearing under these conditions is 125 A. The ratio of duration of growth to the duration of decay of bearing current is taken as 1.4 for transient to steady-state value of bearing current as 0.56^6 and Δ as 0.1. The number of cycles thus obtained matches closely with the duration of the operation of bearing in actual practice (about 2 h).

13.8.2 Lubricant deterioration leading to crater formation on the liner surface of bearings

When the hydrodynamic journal bearing operates under the influence of the shaft voltage, the induced voltage develops on the bearing surface. Under the influence of shaft voltage, two situations arise, depending on the resistivity of the used lubricant. If the resistivity of the lubricant is higher, the bearing acts as a capacitor and energy is stored in the crevice space. If the resistance of the bearings becomes very low when the water contaminated lubricant or low-resistivity lubricant is used, the bearing does not act as a capacitor and the threshold voltage, responsible for allowing an increase in the passage of current, becomes very low. Under these conditions, the energy stored in the inductive–resistive (L-R) circuit of the bearing starts leaking through the liner surface in each rotation. So the growth and decay of the bearing current continue as the shaft rotates and affects, particularly, the liner surface in the zone of minimum film thickness. This leads to the formation of craters due to the passage of high current, initially on the location of high 'points' of the contact and gradually distributed in the zone of minimum film thickness on such area where the bearing resistance is low, as shown in the lower half of the damaged bearing (Fig. 13.1).

The above process increases the formation of craters, and leads to gradual higher bearing vibrations and the failure of bearing in due course by phenomenon of 'electrical pitting'.

13.8.3 Remedial measures

After changing the existing contaminated lubricant with fresh lubricant of higher resistivity, the bearing failure problem did not occur and the bearings operated without any further problem. Since, in this case, crater formation was a surface phenomenon, the craters were removed by scrapping and existing bearings were used by proper cleaning and maintaining the required clearance/diameter ratio.

13.9 Conclusions and recommendations

Based on the investigations concerning diagnosis of bearing problem of synchronous condenser, the following conclusions are drawn:[8]

- The minimum number of cycles for formation of craters of specified volume on the liner surface depend on the bearing dimension, specific load, liner properties and the bearing resistance.
- The minimum number of cycles for the appearance of craters vary inversely to the square of current, bearing inductance, ratio of duration of growth to decay of bearing current and the ratio of asperity to the net contact surface area of the bearing liner.
- The number of cycles for appearance of craters on the bearing surface match closely that found from the actual site data and investigations.
- The cause of the formation of craters on the bearing surface is the low bearing resistance because of water contamination of the oil leading to its very low resistivity, which allows high current to be passed by reducing the threshold voltage to a very low value.
- By using the high-resistivity lubricant and scraping the craters formed, if craters/pitting are limited to the liner surface, the bearings can be reused after complete examination.
- Lubricant characteristics should be properly ascertained before using in the bearings. In particular, water contamination should be avoided.

13.10 References

1 Boyd, J. and Kaufman, H.N., 'The Causes and Control of Electrical Currents in Bearings', *Lub. Eng.*, **15**(1), 28–35, 1988.
2 Chu, P.S.Y. and Cameron, A., 'Flow of Electrical Currents Through Lubricated Contacts', *ASLE Trans.*, **10**, 226–234, 1967.
3 Bradford, M., '*Prediction of Bearing Wear due to Shaft Voltage in Electrical Machines*', ERA Technology Limited, 1990.
4 Prashad, H., 'Effects of Operating Parameters on the Threshold Voltages on Non-insulated Rolling-element Bearings under the Action of Electrical Currents', *Wear*, **117**, 223–240, 1987.
5 Prashad, H. and Rao, K.N., 'Analysis of Capacitive Effect and Life Estimation of

Hydrodynamic Journal Bearings on Repeated Starts and Stops of a Machine operating under the Influence of Shaft Voltage', *Tribol. Trans.*, **37**(3), 641– 645, 1994.
6 Prashad, H., 'Analysis of Inductive Effects of Bearings under the Influence of Shaft Voltages', *BHEL J.*, **15**(1), 26–31, 1994.
7 Starling, G.S., *Electricity and Magnetism*, Green and Co., New York, 1975.
8 Prashad, H., 'Diagnosis of Bearing Problem of Synchronous Condenser–An Experimental and Theoretical Investigation', *Wear*, **188**, 97–101, 1995.

13.11 Nomenclature

C	specific heat of bearing liner ($\mathrm{J\,kg^{-1}K^{-1}}$)
C_d	bearing diameter clearance (m)
d	shaft diameter (m)
E	Young's modulus of elasticity ($\mathrm{Nm^{-2}}$)
H	depth of craters on bearing liner (m)
I_b	bearing current (A)
K	number of craters
L	bearing length (m)
L_j	bearing inductance (H)
n	speed (rpm)
N	minimum shaft revolutions/cycles for appearance of craters on liner surface
N_p	number of cycles to dissipate energy Q_n
p	load per unit area ($\mathrm{Nm^{-2}}$)
Q	electrical energy between journal and bearing (J)
Q_n	electrical energy dissipated to K points between journal and bearing liner in N_p cycles (J)
r	radius of crater (m)
R_s	bearing resistance (Ω)
t_g	time required for growth/increase of bearing current (s)
t_d	time required for decay/decrease of bearing current (s)
T_n	instantaneous rise of high 'points' contact temperature due to energy dissipation (°C)
V	shaft voltage (V)
V_v	net volume of different size craters ($\pi r^2 KH$) ($\mathrm{m^3}$)
W_b	width of contact between the journal and bearing in the zone of load-carrying oil film (m)
ψ	relative bearing clearance (C_d/d)
ρ	density of bearing material ($\mathrm{kg\,m^{-3}}$)
Δ	ratio of asperity to the net contact surface area of bearing liner and journal in each shaft revolution

14
The cause of noise at the top bearings of vertical pump-motor sets

14.1 Introduction

In the case study in question, an irregular, non-periodic, erratic whistling noise was observed quite often at the motor top bearing during the operation of a vertical turbine-type pump. Moreover, a banging noise was observed during starting/stopping of the pump. The pump, of $3000\,m^3 h^{-1}$ discharge capacity, is used to supply raw water in a water works at 45 m head, and is coupled, through a flexible coupling, to a 550 kW squirrel cage induction vertical motor running at 1000 rpm. A duplex bearing at the top of the motor and a roller bearing at the bottom of the motor near the coupling are incorporated (Fig. 14.1). The noise was observed to originate from the duplex bearing for a duration of as little as 2 s to as high as 20 min continuously in a very abrupt pattern, with a periodicity of 10 to 12 days and, sometimes, two or three times a day. To overcome the whistling problem, the clearance in the top duplex bearing was changed. But this proved to be a temporary remedy only, as the noise started reappearing irregularly and non-periodically as mentioned earlier. The banging noise during starting/stopping of the pump was not affected by altering the clearance in the bearing.

Detailed investigations were made, using a systems approach, to discover the probable causes of the noise and to understand the unusual behaviour.

14.2 Sump layout and construction

The following features of sump intake construction and sump layout (Fig. 14.2) were noticed, which could contribute a considerable amount of noise and vibration of the pump by generation of a large number of surface vortices in the sump:

- Zone of separation, and stagnation of water near the intake gates of the sump – causing non-uniformity of flow at the intake to the sump.
- Omission of strainers at the sump intake gate.

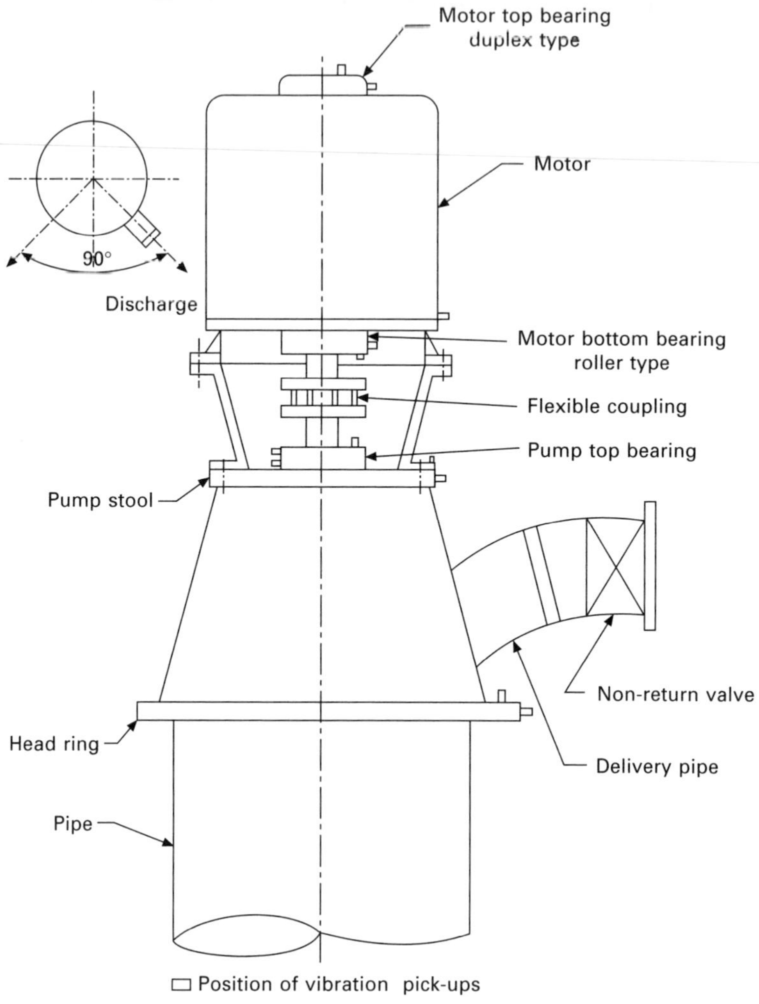

14.1 Pump-motor configuration with pick-ups.

- Network of sharp-edged square columns and beams around the pump intake inside the sump.

14.3 System layout

Figure 14.3 shows the schematic layout of the system. Delivery pipes of the pumps were connected to a common header pipe (of 1.2 m diameter) laid outside the pump house. The delivery pipe diameter for each pump was 600 mm. Reflux (non-return) valve and motor-operated butterfly on–off valve were installed in the delivery pipeline of each pump. The non-return valves fitted on the delivery pipelines of pumps 1 and 2 were equipped with 37 mm

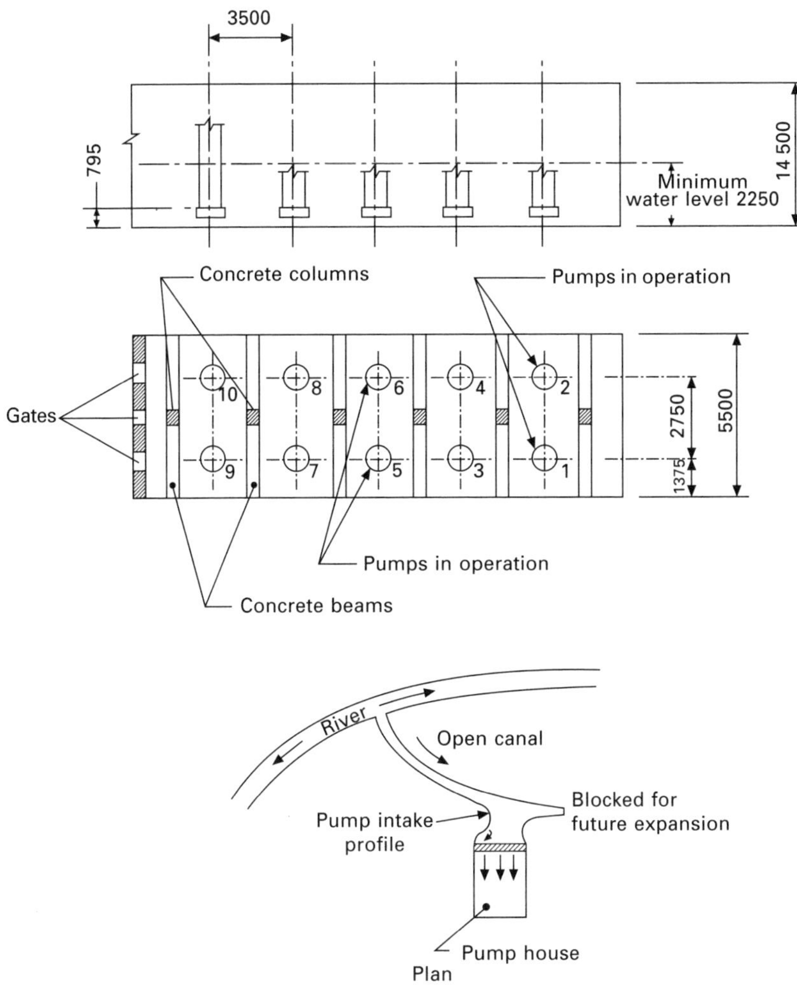

14.2 Sump intake and layout. All dimensions are in mm.

bypass valves of globe type, whereas the non-return valves fitted on the delivery pipelines of pumps 5 and 6 were not equipped with any bypass valves. Moreover, the delivery head of pumps 1 and 2 was 52 m, as against the delivery head of 45 m for pumps 5 and 6 at the discharge design value of $3000\,m^3h^{-1}$. Other pumps shown in the system layout were unoperational during the studies.

14.4 System behaviour

The system behaviour and performance of the pumps were studied under different modes of operation. The following phenomena were observed:

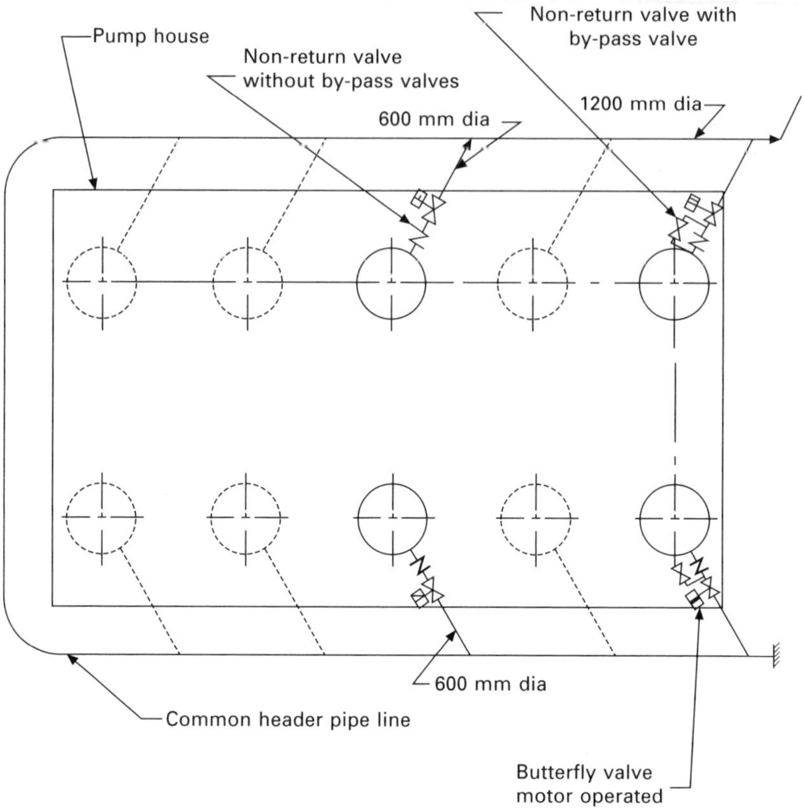

14.3 System layout.

1 On starting pump 6, when all the other pumps in the sump were under operation, it was observed that the swinging flap of the non-return valve, installed on the delivery line of this pump, opened with a heavy bang on the pipeline, resulting in a huge backward impact force on the motor. The deflection of the structure was as high as 3 mm, which could be noticed even with the naked eye. Pump 5 also behaved similarly under these conditions of operation.

2 When all the other pumps were not operating for a duration of about an hour or so, there was no banging noise on starting pump 6, contrary to case (1).

3 On re-starting of pump 6, when the other pumps were just stopped, no intense banging noise due to impact was heard, contrary to case (1).

4 When pump 1 or 2, with its bypass valve in the open position with respect to the non-return valve, was started under the usual start-up conditions, no banging noise was observed.

5 When pump 1 was started by closing the bypass valve with respect to the

non-return valve, to simulate the conditions identical to case (1), under which pump 5 or 6 was started, a similar banging noise and thrust on the motor structure was again observed.

6 An irregular, intermittent, erratic whistling noise was observed at the motor top bearings during operation of pump 5 or 6. No such noise was observed in the case of pump 1 or 2.

14.5 Factors causing the unusual system behaviour

When a pump without a bypass valve is started, with all the other pumps in operation, the pump has to struggle to push open the non-return valve flap instantaneously, resulting in a water hammer on the pump-motor structure. The repeated unusual bangs on the pump-motor structure due to frequent starts and stops of the pump caused a large number of dislocations in various precisely assembled components such as bearings and their housings in the motor, as brought out in cases (1) and (5) in Section 14.4. This phenomenon resulted in the erratic whistling noise at the motor top bearings of pumps 6 and 5. However, in such pumps, the intensity of the banging noise and impact was found to be much less in the absence of a fluctuating delivery head acting on the flap of non-return valve, owing to non-operation of the other pumps in the system, as discussed in cases (2) and (3) in Section 14.4. The presence of a bypass valve across the non-return valve dampened the pressure oscillations coming onto the valve flap from downstream of the pump and, thus, the unusual bangs on the structure were obviated, as in case (4) in Section 14.4. Hence, no whistling noise from the motor top bearings of pump 1 or 2 was observed.

It was also noted that the air vent valve, installed near the non-return valve, facilitates the momentary opening of the non-return valve by venting the air entrapped in the pump.

14.6 Vibration spectra and analysis

The vibration frequency spectra for different conditions of operation of the pump-motor sets were recorded on bearing housings of the pump (top), the motor (top and bottom), pipe discharge head ring, and the pump bearing stool, as per the location of pick-ups shown in Fig. 14.1; vibration data are given in Tables 14.1 to 14.5. The pick-ups were mounted in line with the discharge pipe, and at an angle of 90° with respect to it in both horizontal and vertical directions, and under 'load' as well as 'no-load' conditions. The intake screen of the pump 6 was blocked by a flat plate of size (500×500) mm^2 and the resulting vibration levels of pump top and motor bottom bearings were studied. The behaviour of pump 6 was investigated and the vibration spectra of pump and motor bearings analysed under the conditions when the

Table 14.1 Axial and radial vibrations at various frequencies of pump bearing

(a) Axial

Serial no.	Frequency (Hz)	Vibration levels under full load		Vibration levels with other machines stopped and gates fully opened		Vibration levels with other machines stopped and gates fully closed	
		ms^{-2}	μm	ms^{-2}	μm	ms^{-2}	μm
1	16	0.18	17.84	0.31	30.7	0.35	34.7
2	64						
3	116	0.35	0.66	0.28	0.53	0.43	0.8
4	1056	0.088	0.0019				
5	1156–1184	0.0588	0.0012	0.09	0.0016	0.13	0.0024
6	1656						

(b) Radial

Frequency (Hz)	Vibration levels under full load and pick-up in line with discharge		Vibration levels under full load and pick-up 90° away from discharge		Vibration levels under full load with other machines stopped and gates fully opened with pick-up in line with discharge		Vibration levels under full load with machines stopped and gate fully closed with pick-up in line with discharge	
	ms^{-2}	μm	ms^{-2}	μm	ms^{-2}	μm	ms^{-2}	μm
16	0.98	97	0.676	67	0.7	69.3	0.75	74
64	0.2	1.23	0.15	0.93				
116	0.77	1.45	0.6	1.13	0.58	1.1	0.56	1.1
1056								
1156–1184	0.078	0.00148	0.078	0.00148	0.08	0.0015	0.08	0.0015
1656	0.078	0.00072	0.078	0.00072				

Overall axial vibration levels = 30 μm.
Overall radial vibration levels = 70 μm.

pump operated alone with sump gates fully opened and under closed conditions. The frequency spectra in the range of 0–2000 Hz were analysed, and recorded on a graphics level recorder. Instruments of type M/s B & K were used for vibration analysis. Some of the typical plots are shown in Fig. 14.4 to 14.9 for the frequency spectra of the motor top and bottom bearings and stool of the pump bearing with the axial as well as radial directions, respectively. The shock-pulse meter technique was also used to assess the condition of the bearings. The vibration analysis of pump-motor sets 5 and 6 only is discussed here, as pump-motor sets 1 and 2 did not exhibit any noise problem, because

Table 14.2 Axial and radial vibration levels at various frequencies of motor top bearing

(a) Axial

Frequency (Hz)	Vibration levels under full 'load'		Vibration levels under 'no-load'	
	ms^{-2}	μm	ms^{-2}	μm
16	0.24	23.8	0.013	1.29
116	0.11	0.21	0.003	0.0056
312	0.25	0.065		
1156–1168	6.31 Fig. 14.4	0.12	0.016	0.0003

(b) Radial

Frequency (Hz)	Vibration levels under full 'load'				Vibration levels under 'no-load'	
	Pick-up in line with discharge		Pick-up 90° away from discharge			
	ms^{-2}	μm	ms^{-2}	μm	ms^{-2}	μm
16	0.92	91	0.31	30.7	0.011	1.09
116	0.17	0.34	0.36	0.68	0.0055	0.011
312						
1156	4.2 Fig. 14.5	0.08	11.33	0.22	0.04	7.4×10^{-4}

Overall axial vibration levels = 50 μm.
Overall radial vibration levels = 50 μm.

of incorporation of the bypass valve and air vent valve, and adequate slenderness ratio of the intermediate structure between pump and motor.

14.7 Results and discussion

14.7.1 At 16 Hz frequency and shock pulse values

1 The d_b values, as shown by the shock pulse meter, were found to range between 10 and 15, indicating that all the bearings of the system were normal.
2 Under coupled conditions, the vibration levels in the horizontal direction at the running speed frequency (16 Hz), with the pick-ups mounted in line with the discharge pipe, were 97, 25.7 and 91 μm as against 67, 21.2 and 30.7 μm with the pick-ups mounted at an angle of 90° with respect to the discharge pipe, at pump bearing, motor bottom bearing and motor top bearing, respectively (Tables 14.1 to 14.3). The excessive vibration levels in the former case were attributed to excitation due to pressure head and high flow velocity along with the pressure fluctuations in the

Table 14.3 Vibration levels at various frequencies of motor bottom bearing

(a) Axial

Frequency (Hz)	Vibration levels under full 'load'		Vibration levels under 'no-load'	
	ms^{-2}	μm	ms^{-2}	μm
16	0.394	39.0	0.45	44.5
64				
116	1.274	2.4	0.38	0.72
1056–1064	1.51	0.0338		
1156–1168	5.06	0.10	0.66	0.012

(b) Radial

Frequency (Hz)	Vibration levels under full 'load'				Vibration levels under 'no-load'	
	Pick-up in line with discharge		Pick-up 90° away from discharge			
	ms^{-2}	μm	ms^{-2}	μm	ms^{-2}	μm
16	0.26	25.7	0.214	21.2	0.27	26.7
64	0.13	0.81	0.108	0.67	0.2	1.2
116	0.05	0.09	0.125	0.235	0.05	0.09
1056–1064	0.32	0.0073	0.64	0.014		
1156–1168	0.44	0.008	4.47	0.084	0.26	0.005

Overall axial vibration levels = 25 μm.
Overall radial vibration levels = 40 μm.

Table 14.4 Axial and radial vibration levels at various frequencies at pipe discharge head ring

	Axial		Radial			
Frequency Hz	Vibration levels under full 'load'		Vibration levels under full 'load'			
			Pick-up in line with discharge		Pick-up 90° away from discharge	
	ms^{-2}	μm	ms^{-2}	μm	ms^{-2}	μm
16	0.174	17.2	0.2	19.8	0.18	17.8
116	0.45	0.85	0.333	0.63	0.09	0.17
132	0.22	0.32			0.1	0.15
1156	0.07	0.0013	0.11	0.002	0.07	0.0013

Table 14.5 Axial and radial vibration levels at various frequencies at pipe bearing stool

Frequency Hz	Axial		Radial			
	Vibration levels under full 'load'		Vibration levels under full 'load'			
			Pick-up in line with discharge		Pick-up 90° away from discharge	
	ms^{-2}	μm	ms^{-2}	μm	ms^{-2}	μm
16	0.22	21.8	0.5	49.5	0.27	26.7
112–116	0.23	0.46	0.45	0.90	0.38	0.72
196	0.27	0.22				
1152–1156	1.0	0.019	0.8	0.015	0.53	0.01

Overall axial vibrations levels = 120 μm.
Overall radial vibration levels = 50 μm

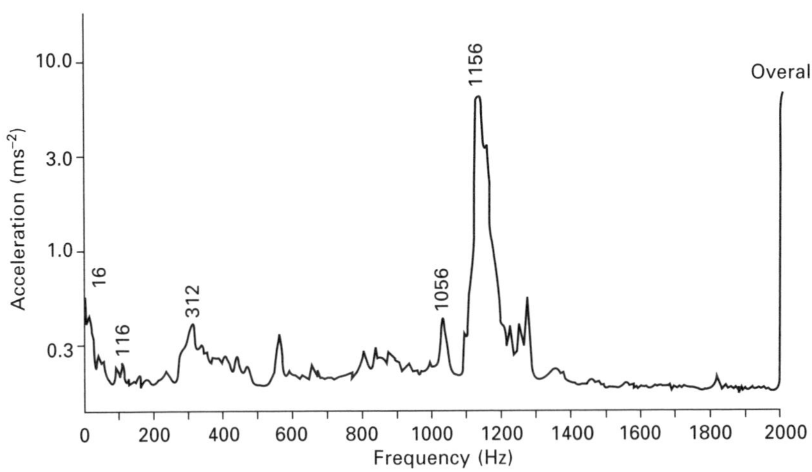

14.4 Frequency spectrum of motor top bearing under full load with pick-up mounted on bearing housing in axial direction.

discharge pipeline. Weak support and high slenderness ratio of the intermediate structure between pump and motor were also contributory factors. The same holds good at the pipe discharge head ring and at the pump-bearing stool. At these locations, the horizontal vibration levels, in line with the discharge pipe, were 19.8 and 17.8 μm, whereas the corresponding values at an angle of 90° with respect to the discharge pipe were 49.5 and 26.7 μm (Tables 14.4 and 14.5).

3 The higher vibration levels in the horizontal direction at the pump bearing as against the motor bearings, as shown in (2), also indicates higher imbalance in the pump than in the motor shaft. The insignificant value

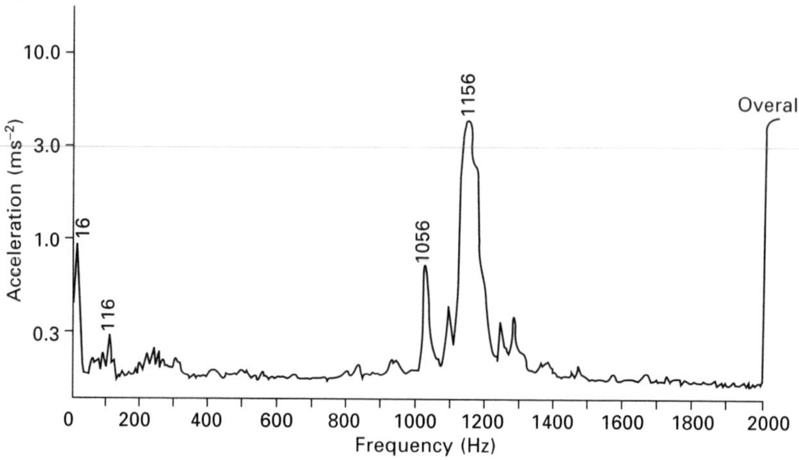

14.5 Frequency spectrum of motor top bearing under full load with pick-up mounted in line with discharge on bearing housing in radial direction.

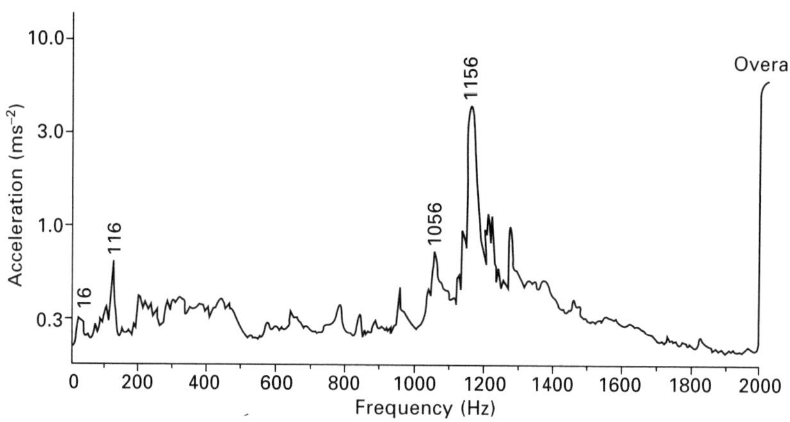

14.6 Frequency spectrum of motor bottom bearing under full load with pick-up mounted on bearing housing in axial direction.

of the horizontal vibration level (1.09 μm) at the motor top bearing, under uncoupled conditions, indicated that the motor is well balanced (Table 14.2). However, the higher value of horizontal vibration level (26.7 μm) at the motor bottom bearing was due to the overhang of the motor shaft, as the motor was decoupled from the pump under these conditions (Table 14.3).

4 The frequency components at 64 and 116 Hz were observed because of the harmonics of running speed frequency and the rolling-elements in the pump and motor bearings (Tables 14.1 to 14.3).

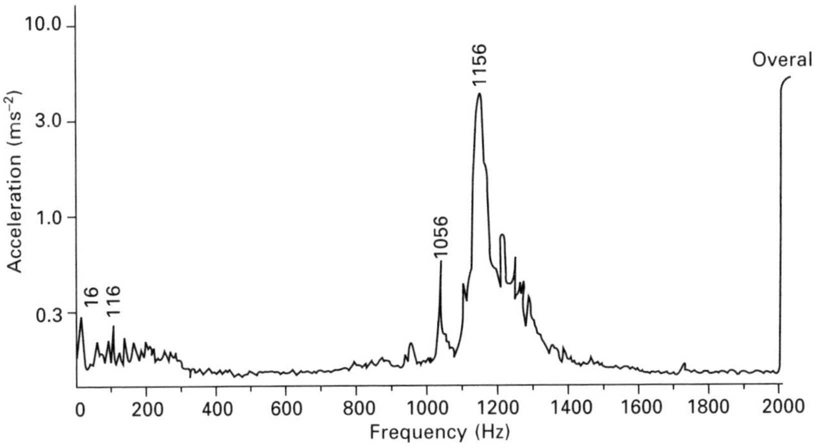

14.7 Frequency spectrum of motor bottom bearing under full load with pick-up mounted 90° away from discharge on bearing housing in radial direction.

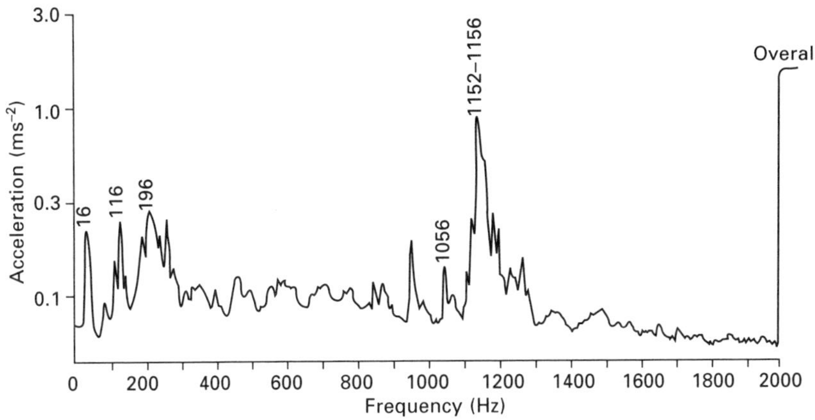

14.8 Frequency spectrum of pump bearing stool under full load with pick-up mounted in axial direction.

5 Reduction in horizontal/radial vibration levels from 97 to 69.3 μm at 16 Hz when the other pumps were stopped, was attributed to the effect of parallel operation of the pump in the same sump and also to the influence of the pressure head, as well as flow fluctuations in the discharge pipe on the start-up characteristics of the pump (Table 14.1).

6 The increasing order of vibration levels in axial directions at 16 Hz, from the pipe discharge head (17.2 μm) to the pump bearing stool (21.8 μm) and to the motor bottom bearing (39 μm), is a clear indication that the flexible coupling was not capable of absorbing the axial thrust and that

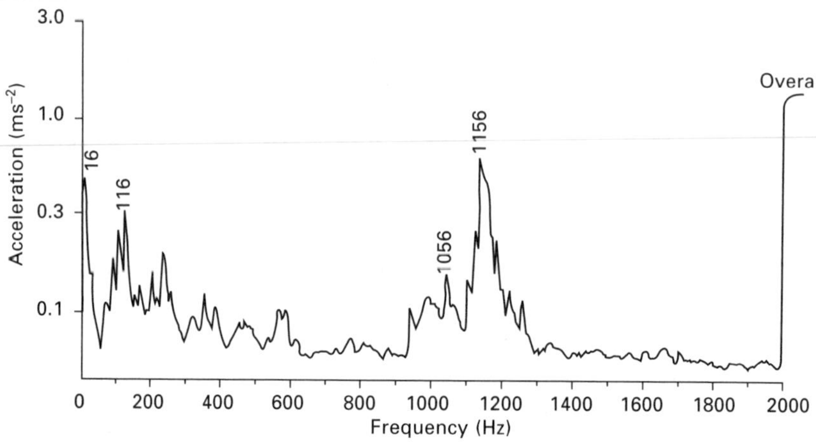

14.9 Frequency spectrum of pump bearing stool under full load with pick-up mounted in line with discharge in radial direction.

the thrust was getting transmitted to the motor bearing from the pump side (Tables 14.3 to 14.5).

7 The higher overall vibration level in the horizontal direction (70 μm), compared with that in the axial direction (30 μm), at the pump bottom bearing, and similarly at the motor bottom bearing (40 μm in the horizontal direction and 30 μm in the axial direction) as shown in Tables 14.1 and 14.2 indicates that the intermediate structure between pump and motor requires strengthening in the horizontal/radial direction. It also indicates that the slenderness ratio of the intermediate structure was higher than the required value.

8 A higher overall vibration level in the axial direction at the pump bearing stool (120 μm) as shown in Table 14.5 indicates that vibration isolation should be provided between the pump-stool and the intermediate structure, to eliminate the transmission of axial vibration towards the motor.

9 An insignificant change in the vibration levels at the delivery pipe on blocking the intake screen, indicates that the partial blocking has no effect on the performance of the system.

10 Under the conditions when all the machines in the sump were stopped and the gate was gradually closed to reduce water level in the sump, the axial and radial pump vibration levels have increased from 30.7 to 34.7 μm and 69.3 to 74 μm, respectively. The radial vibration levels were even more than that recorded by the pick-up mounted at a point 90° away from the discharge as shown in Table 14.1. This indicates that a certain minimum level must be maintained in the sump, otherwise at the suction of the pump, considerable noise and vibration will be generated.

14.7.2 At 1156 Hz frequency

The frequency component at 1156 to 1184 Hz at the bearing housing and pump bearing stool as indicated in Fig. 14.4 to 14.9 were attributed to the presence of resonance frequency of the structure. However, displacement levels at this frequency were significantly low (Tables 14.1 to 14.5), but higher acceleration levels at the bottom and top motor bearings and pump bearing stool were very dangerous owing on the forces exerted to the system at this frequency.

The hydraulic forces affected the rigidity of the structure near the discharge. That is why at 1156 Hz the pump bearing stool showed 1.0 ms^{-2} acceleration levels in the axial direction rather than that of 0.07 ms^{-2} acceleration levels at discharge head ring. Similarly, acceleration levels in the radial direction in line and 90° away from the discharge at the pump bearing stool were 0.8 and 0.53 ms^{-2} rather than 0.11 and 0.07 ms^{-2} at the pipe discharge head ring, respectively (Tables 14.3 and 14.4). This difference is attributed to the fact that the head ring was rigidly fixed and much larger than the bearing stool.

At any point after discharge, the hydraulic forces did not affect rigidity. This pattern is more evident at the motor top and bottom bearings, which were located away from the discharge pipe as compared with the pump bearing stool and the discharge head ring. The radial acceleration levels at the top bearing were 11.33 and 4.2 ms^{-2} at 90° away and in line with the discharge as against 4.47 and 0.44 ms^{-2} on the bottom bearing, respectively. Similarly, in the axial direction acceleration levels were 6.31 and 5.03 ms^{-2} at the top and bottom bearings respectively (Fig. 14.4 to 14.7 and Tables 14.2 and 14.3). At the natural frequency of the structure, acceleration levels at a point away from discharge were not influenced by hydraulic disturbances and were higher in comparison to the point near to and in line with discharge. It may also be corroborated from the overall vibration levels of 50 μm both in the radial and axial directions at the motor top bearing (Tables 14.2 and 14.3) and may be attributed to the cantilever design of motor supporting structure.

The difference between the acceleration levels in line and 90° away from the discharge at the motor top and bottom bearings may be related to the difference in the dynamic stiffness in these directions at 1156 Hz.

14.8 Design of the bearing used and its significance

The duplex bearing used at the top end of the motor took the axial load in a downward direction and momentary upward thrust during the start of the pump. The duplex bearing was a single row angular contact ball bearing, which had the raceway so arranged that the bearing could carry axial load in both directions. This bearing had a two-piece inner ring provided with a number of balls. Thus, it ensured a high load-carrying capacity and operated best under predominantly axial loading. Because of this design, the outer ring and ball assembly could be mounted independently on the inner ring.

14.9 Explanation of the cause of noise at the motor top bearing

Noise in rolling-element bearings is quite common. Insufficient operating clearance in the bearing may cause an uncommon running noise or a whistling or singing noise in the rolling-element bearing. Low-pitched rumbling or irregular noise in the rolling-element bearing may be due to excessive operating clearance or due to damaged running surfaces.

Since the cause of the appearance/disappearance of irregular, non-periodic, erratic whistling noise from the motor top bearing could not be directly correlated with the general causes of noise in bearings, the same may be explained in the light of the vibration analysis (see Sections 14.4 and Section 14.6, and the discussions in Section 14.7.1).

Since raw water was directly drawn from the river (Fig. 14.2), foreign particles were liable to be sucked in by the pump owing to the absence of screens at the main gate and the presence of a large number of surface vortices in the sump. Whenever any foreign particle gets into the impeller running clearance, the pump impeller is lifted. Because the flexible coupling is not axially free, as shown in point (6) in Section 14.7.1, this lift is transmitted to the rotor of the motor from the bowl assembly, as shown in Fig. 14.10. As soon as the rotor of the motor was lifted, this released/balanced the vertical downward load of the motor at the top duplex bearing. Under these conditions, the bearing operated practically under 'no-load' in unstable conditions. Owing to the 'slip' phenomenon occurring under 'no-load' operation, noise is expected to be generated because of skidding of the rolling elements. The duration of the nose would depend upon the extent of the foreign particles remaining entangled with the impeller. Dislocations in the precise bearing components by unusual bangs on the motor structure on repeatedly starting and stopping of the pump, in the absence of the bypass valve, also add to the noise.

14.10 Conclusions and recommendations

On the basis of the analysis of the problem, the following recommendations were made:

- Noise and vibration can be considerably reduced by providing screens at the intake of the sump and smoothing the network of sharp-edged square columns and beams around the pump intake inside the sump, to prevent the formation of vortices, which leads to non-uniformity of flow.

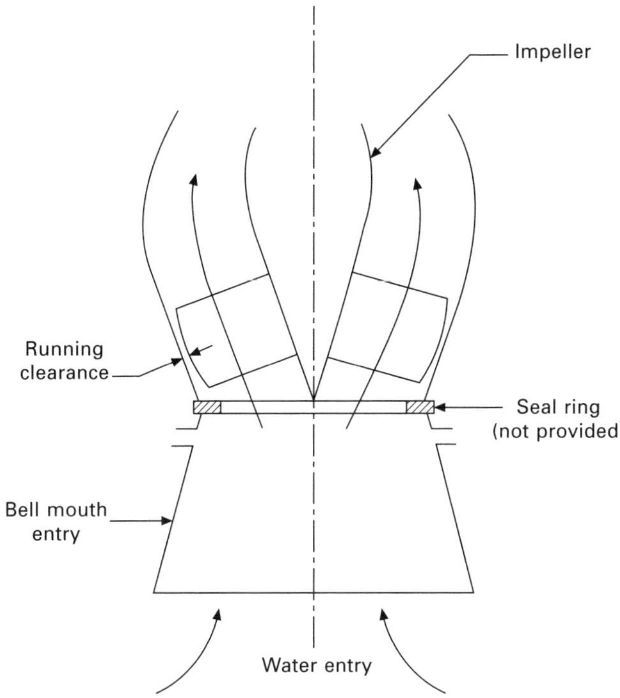

14.10 Typical constructional feature of the bowl assembly.

- The non-return valve is to be equipped with a bypass valve, to avert the heavy bangs of the system.
- The air vent valve should be installed very near the non-return valve, so as to vent the air entrapped in the pump in order to facilitate the momentary opening of the non-return valve.
- The slenderness ratio of the interconnecting structure between pump and motor should be reduced and the support should be strengthened to avoid higher vibration levels and minimize the deflection of the support during the start-up conditions.
- Vibration isolation should be provided between pump-stool and the intermediate structure.
- The coupling used should be truly flexible so as to absorb the complete axial thrust.

The above recommendations have been implemented, and the system has been working satisfactorily since then.

14.11 Bibliography

1 *Pump Handbook*, edited by Karassik, J., Kritzsch, W.C., Fraser, W.H. and Messina, J.P., McGraw-Hill, New York, 1976.
2 *Pump Application Engineering*, Hicks, T.G. and Edwards, T.W. McGraw-Hill, New York, 1971.
3 International Standard ISO 2372 and ISO 3945-1977(8).
4 Prashad, H., Chauhan, R.B.S., Panwalkar, A.S. and Narsimhan, G.S.L., *BHEL, Feedback J.*, 18–24, June 1987.

15

Modifications to the design and bearings of horizontal axis windmills used for pumping water, to achieve trouble-free, reliable operation

15.1 A general review

In this chapter, the operational philosophy and performance of a few horizontal-axis windmills, used for pumping irrigation and drinking water, are discussed. The role of bearings and their effect on the overall performance of the windmills have been investigated and it has been found that the operational performance of the windmills is closely related to the behaviour of the bearings of the windmills. The ultimate effect of the bearing performance on the functioning of single-acting reciprocating pumps has been identified.

The behaviour of the bearings of the windmills under the effect of the forces acting on them and the subsequent effect of this behaviour on other components of the windmills was studied. The investigations cover the cause of the reciprocating pumps not functioning in spite of the rotor of the windmills in rotation. Based on this, three recommendations have been made to ensure trouble-free and reliable operation of the windmills and reciprocating pumps under various operating conditions: replacement of the bearings used in the swivel box; replacement of the windmill rotor bearings; and modification of the existing joints between the pump-rods. The modification of the existing joints between the pump-rods was effected and the results were satisfactory.

15.2 Introduction

The search for alternative energy sources has led to the re-discovery of wind power. Wind energy can play a major role in water pumping for irrigation and drinking purposes and, hence, can make a significant contribution to rural development worldwide. The cumulative impact of the effective use of small amounts of energy in rural areas can be considerable and, for developing countries, it could make a significant contribution in increasing the per capita energy available, a prerequisite for improving the quality of life, economic growth and social well-being. Agricultural productivity, public health and communications would all benefit from the availability of a low-power decentralized system using wind energy.

Investigations were carried out on a few windmills (used for pumping water), to determine the nexus between the bearings of the windmills and the functioning of the reciprocating pumps. This study details the investigation undertaken and the recommendations made to ensure trouble-free and reliable operation of the windmills and the reciprocating pumps.

15.3 Design features

The major components of the windmills used for pumping water were tower, rotor, transmission system, tail and auxiliary vanes, and pump. Windmills were of an all-metal construction, with low cut-in wind speed, and were designed with optimum aerodynamic design of blades for long life, trouble-free operation and minimum maintenance. To ensure the safety of the windmills, an automatic furling mechanism was provided to take care of high wind speeds. The windmills had 7 m high towers and were three-legged tubular structures. The tower was designed to withstand a wind speed of 160km h^{-1}. The rotor (of 5 m diameter) had 12 aerodynamically designed, tapered and twisted sheet metal blades. The power-shaft, of 50 mm diameter, attached to the rotor hub, was supported on a turntable through two deep-groove ball bearings.

The principal technical details of the windmills were as follows:

- rotor blade solidity: 0.32;
- rotor velocity ratio: 2.0;
- designed wind speed: 14 km h^{-1};
- cut-in wind speed: 8 km h^{-1};
- cutout wind speed: 40 km h^{-1}.

15.4 Operational philosophy

As the wind direction changes from time to time, a suitable arrangement was required to ensure that the rotor faced the wind continuously, to extract the maximum possible energy. The turntable made up of framed angle sections was provided for this purpose. A pipe welded at the top of the tower guided it.

The tail plane and side plane were also fixed to the turntable. These were interconnected such that in heavy wind they are automatically disconnected, allowing the turntable to rotate and align the rotor to the wind direction, thus protecting it from strong winds.[1]

15.5 Transmission system

The developed shaft power was transmitted to a reciprocating pump through a transmission system consisting of crank, crank lever, swivel box, connecting rod, reciprocating pipe and pump-rod. The reciprocating pipe was guided inside two cast iron bushes, supported inside the pipe welded to the tower. The lower end extension of the reciprocating pipe was connected to a single-acting reciprocating pump through the pump-rods; the pump, submersible in water, was used for pumping water. A swivel box was used in the transmission, between the connecting rod and the reciprocating pipe. This provided rotary, swinging and reciprocating motion to the connecting rod at the upper end, but only reciprocating motion to the reciprocating pipe at the lower end. The details of the windmills are shown in Fig. 15.1.

15.6 Bearings and performance of the reciprocating pump

The power-shaft, which was attached to the rotor hub, was supported on the turntable through two deep-groove ball bearings of type 63102 ZR or type 62102 ZR. The swivel box used two single-acting thrust ball bearings of type 51103. These bearings were fitted so that the upper race of the top bearing could rotate as the turntable rotated depending on the wind direction, but this rotation was transmitted to the pump-rod. Only reciprocating motion was transmitted to the reciprocating pipe, as the lower race of the thrust ball bearing (type 51103) had a larger inner diameter than the upper race.

However, after a prolonged operation of a number of windmills under various operating conditions, the following phenomena were observed:

- The reciprocating pipe, in addition to transmitting the reciprocating motion, had a tendency to rotate.
- The above tendency forced the pump-rod of the reciprocating pump to rotate.
- The rotation of the pump-rod tended to unscrew and disengage the leather washers in the pump cylinder, in addition to unscrewing the lock nuts and the connecting cylindrical sleeves used at the junction of the connecting pump-rods. Details of the joint between two pump-rods are shown in Fig. 15.2.
- The pump became non-functional when the leather washer was completely unscrewed, although the rotor of the windmill would still be rotating.
- Finally, no water flowed from the reciprocating pump.

A method to stop the rotation of the pump-rod was tried, by providing a slot in the reciprocating pipe and by fixing a bolt to the tower pipe such that the slot would move around the bolt. However, this method was not found to be

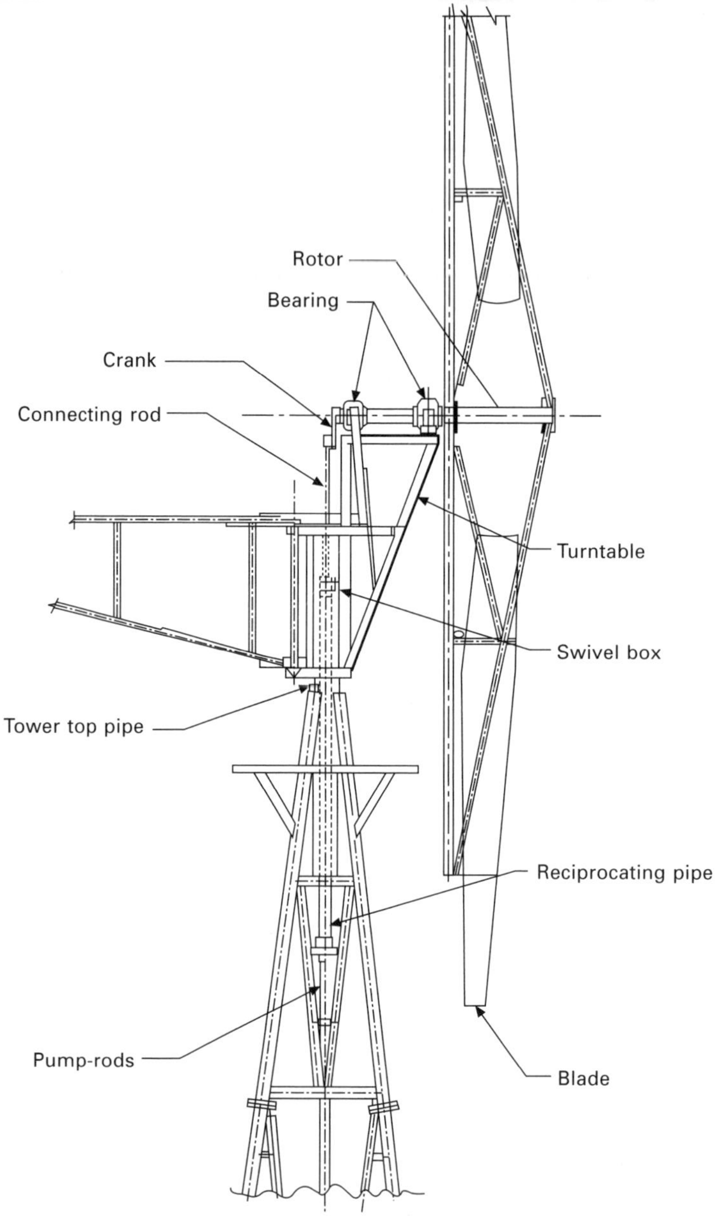

15.1 Horizontal-axis windmill.

an effective solution, as the bolt would get cut within a short period, owing to friction from the reciprocating motion of the pipe.

The nexus between the bearings and the performance of the reciprocating pump was then examined. The effect of the joint between the pump-rods on the performance of the reciprocating pump was also examined.

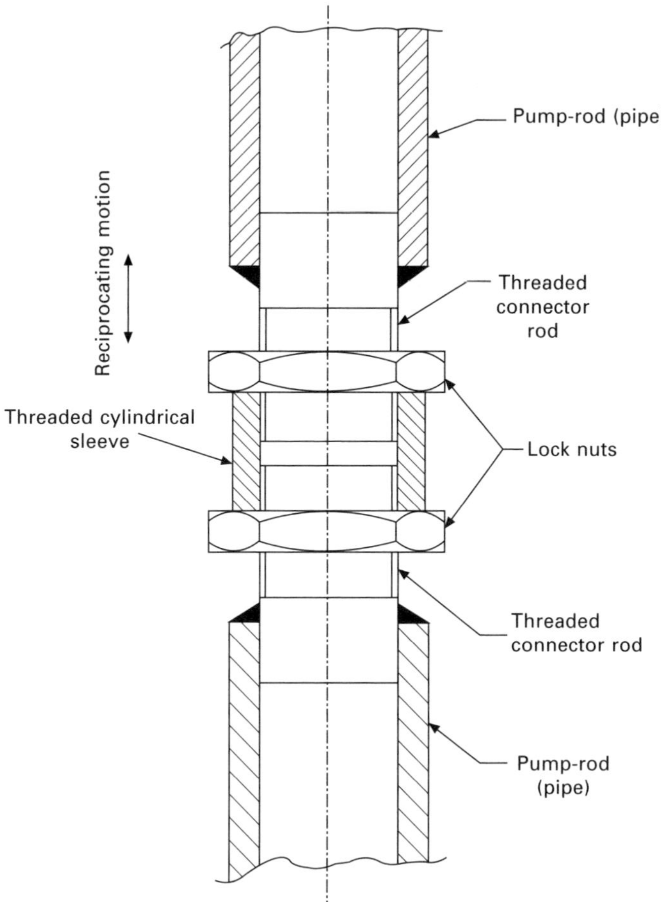

15.2 Details of joints between two pump-rods.

15.6.1 Influence of bearing behaviour on the functioning of the reciprocating pump

Deep-groove ball bearings, of types 63102 ZR and 62102 ZR, which were used to support the power-shafts of windmills, were analysed under continuous radial and axial loads of 510 kgf and 100 kgf, respectively. These values of loads were obtained from knowledge of the various forces the windmill was subjected to under the most rigorous conditions of operation. The analysis showed that the deep-groove ball bearings were quite safe and reliable, and have an expected operating life of more than 20 years at 40 rpm speed of the windmill.

The two single-acting thrust bearings of type 51103, used in the swivel box, were also analysed under a continuous axial load of 75.8 kgf and a radial load of 14.6 kgf. These are the maximum calculated forces experienced by the thrust bearings during the rotation of the turntable, based on a particular wind direction. From the analysis, it was found that a minimum axial load of 0.5×10^{-3} kgf was required on the bearings for their normal operation without any pre-loading, but they cannot take any radial load, as they tend to become misaligned. Figures 15.3 and 15.4 show the thrust bearings under axial load and combined axial and radial loads, respectively.

As the wind direction changed, the rotor moved around the reciprocating pipe and the thrust bearings were exposed to combined axial and radial loads, which led to the misalignment in the bearings. Owing to the misalignment in the bearings, there was an increase in the friction between the reciprocating pipe and the lower race of upper thrust bearing and also between the reciprocating pipe and the upper race of the lower thrust bearing. As a result, the reciprocating pipe had a tendency to rotate. The misalignment also generated noise in the bearings.

The bolt used to restrict the rotation (as explained in Section 15.5), was cut on the opposite faces, thereby unscrewing the pump washer. The pump, thus, becomes non-functional.

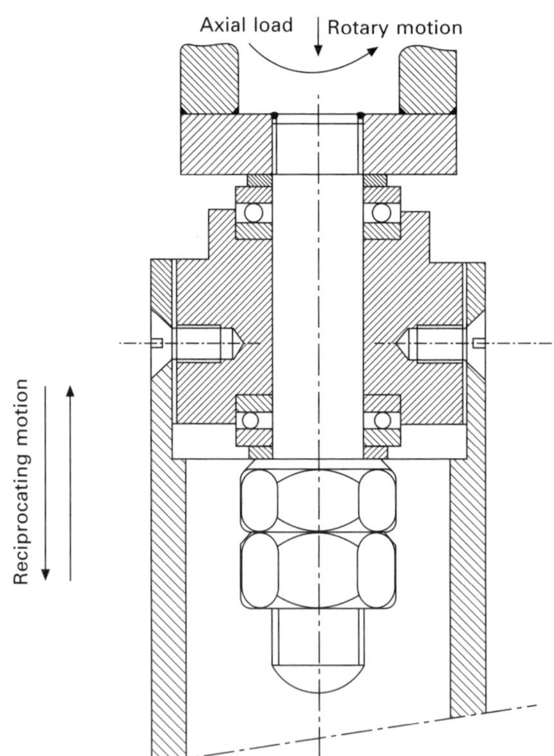

15.3 Thrust ball bearing (type 51103) in swivel box under axial load.

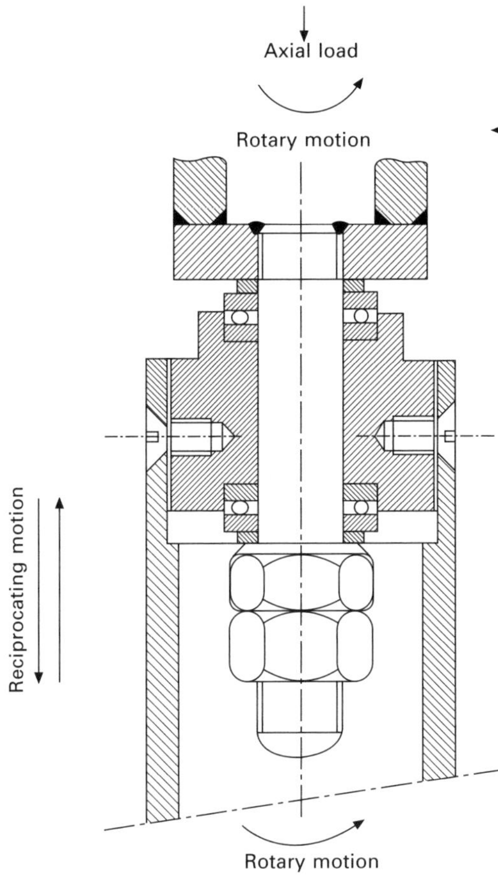

15.4 Thrust ball bearing (type 51103) in swivel box under combined axial and radial loads.

15.7 Recommendations

On the basis of the analysis made, the following three recommendations were suggested:[3]

- replacement of bearings in the swivel box;
- replacement of the windmill rotor bearings;
- modification of pump-rod joints.

15.7.1 Replacement of bearings in the swivel box

Considering that both radial and axial loads act on the bearings in the swivel box, an angular contact thrust ball bearing, of type 234406 MASP, was recommended in place of the two thrust bearings of type 51103. The mounting arrangement of the angular contact thrust ball bearing is shown in Fig. 15.5.

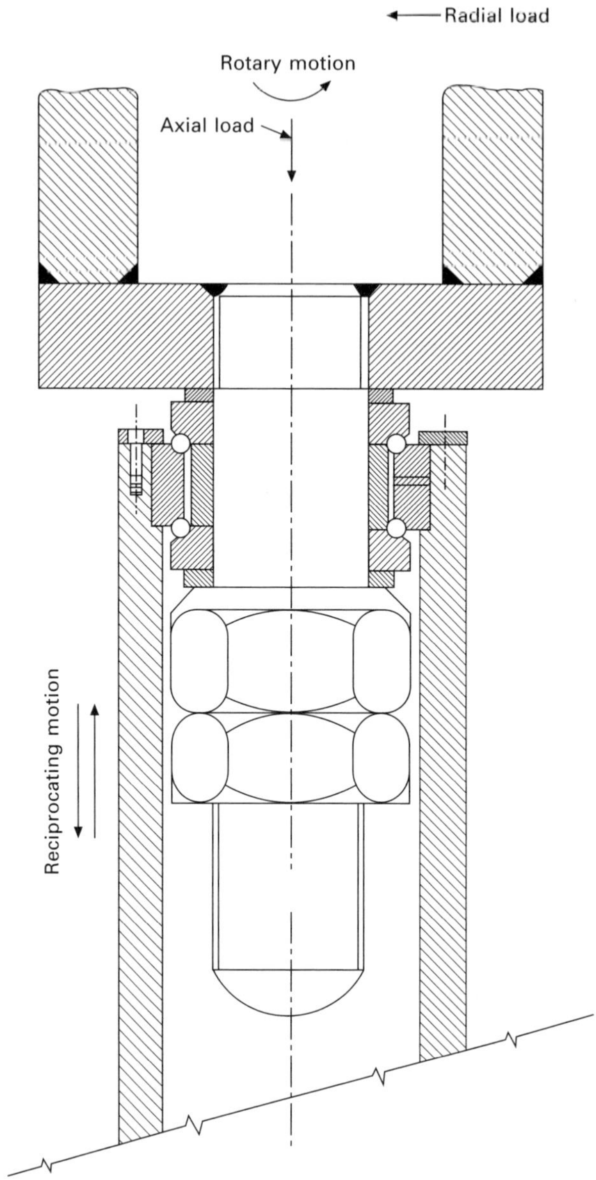

15.5 Angular contact thrust ball bearing (type 234406 MASP) – under combined axial and radial loads.

This bearing can take both axial and radial loads, thus minimizing the rotation of the reciprocating pipe and transmitting only the reciprocating motion to the pump-rod, apart from eliminating the bearing noise generated due to any misalignment.

For an axial load of 75.8 kgf and a radial load of 14.6 kgf, this bearing should have a normal working life of more than 20 years.

15.7.2 Replacement of the windmill rotor bearings

For the windmill to operate under high wind speeds, more analyses were made and it was found that the deep-groove ball bearings (of types 63102 ZR and 62102 ZR) were safe and reliable even for occasional momentary axial and radial load of 1100 and 510 kgf, respectively. However, it was felt necessary that for extra safety and reliability of the bearing, suitable additional locking should be used, to counteract the occasional momentary forces under cyclonic conditions.

For the newly developed series of windmills, keeping in view high reliability and safety, it was recommended that the existing bearings be replaced by a double-row tapered roller bearing at the locating end and a cylindrical roller bearing at the floating end of the windmill rotor. Figure 15.6(a) shows a tapered roller bearing (matched type 3130 AK 11 A 60 100) for supporting the rotor at the locating end, while Fig. 15.6(b) shows a cylindrical roller bearing (type NU 1010 M) for supporting the rotor at the floating end, for a power-shaft of 50 mm diameter. Using these bearings, windmills can be designed and operated at wind speeds ranging from 6.3 to 24 ms^{-1}.[2] The final

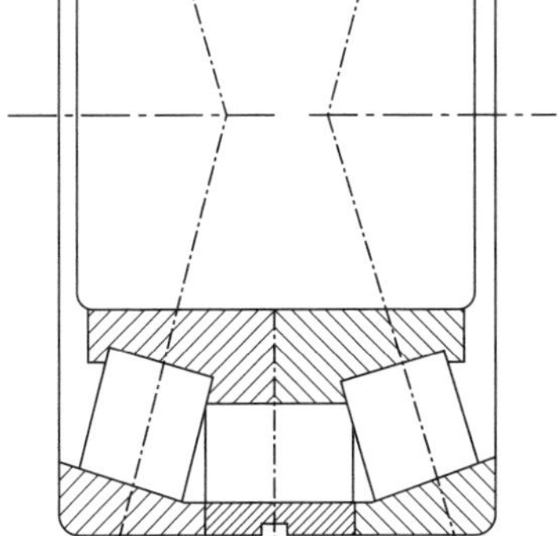

(a) Tapered roller bearing (type 3130 AK 11A 60100) at the locating end

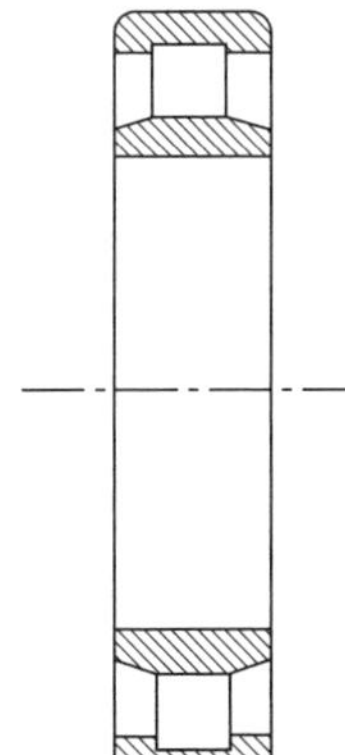

(b) Cylindrical roller bearing (type NU 1010 M) at the floating end

15.6 The bearings recommended for the rotor of the windmill.

calculations in respect of the rotor bearings are to be made on the basis of the different collective loads and their corresponding operating time.

15.7.3 Modifications of the pump-rod joints

To obviate the unscrewing of lock nuts from the threaded connector rod, and loosening of the threaded cylindrical sleeve (by the rotary motion of the pump-rod, resulting in the disengagement of leather washers in pump cylinders), welding of the threaded cylindrical sleeve directly with the threaded connector rod at the upper end, as shown in Fig. 15.7, was recommended. In the modified design of the joint between the two pump-rods, the upper lock nut (as shown in Fig. 15.2) has been removed. The lower lock nut has been retained for dismantling of the joint. In addition, a tail washer has also been introduced on the lock nut to obviate its unscrewing tendency.

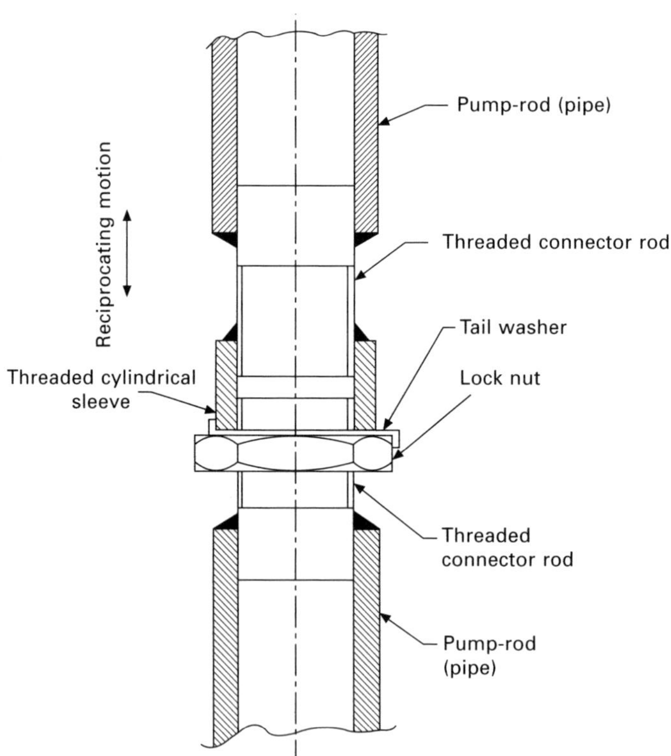

15.7 Modified design of joint between two pump-rods.

15.8 Conclusions

The recommended modification of the pump-rod joints has been effected on a number of windmills. The reciprocating pumps of the windmills at various other sites have been working satisfactorily since the modification was carried out. However, to ensure a trouble-free and reliable operation of the windmills and reciprocating pumps, the bearings in the swivel box should be replaced, and the recommended bearings are to be incorporated in place of existing bearings in the rotor of the windmill. The windmills can then be expected to pump water satisfactorily for more than twenty years.

15.9 References

1 Gupta, R.P., Mantrawadi, S.C. and Chandra, S.K., *Pumping of Water by a Windmill*, BHEL R&D Report, 1985.
2 'Smooth Operation and High Degree of Safety, Large Rolling Bearings in the Wind Power Station GROWAN', *Ball Roller Bearing Eng.* (1) 1983.
3 Prashad, H. and Gupta, R.P., 'Bearings and Design Modifications for Trouble-free and Reliable Operation of Horizontal-axis Windmills used for PumpingWater', *BHEL J.*, **7**(1), 14–19, 1986.

16

Magnetic suspension bearings for AC energy meters

16.1 A general review

Energy meters with conventional or jewel bearings have a limited life span. This is because the jewel bearings tend to become 'sticky' after a few years of operation, resulting in increased frictional torque. Replacing the conventional bearings by magnetic bearings, in which the rotor weight is supported by the magnetic force and contact friction is kept to a minimum, can considerably increase the life span of an energy meter.

This chapter describes the principles governing the design and manufacture of different types of magnetic bearing. The frictional performance of developed magnetic bearings has been compared with those of the conventional type. The comparison brings out the importance of refinement that is necessary in the design and manufacturing of magnetic bearings.

16.2 Introduction

A watt-hour or energy meter in electricity supply systems is essentially an integrating and registering device in which the time integration of the load or power for a given duration is performed by the rotation of an aluminium disc. The optimum accuracy of the rotation of the disc, throughout the serviceable life of the meter, is, therefore, most important. This requirement is directly related to the quality and characteristics of the bearings used in the meter to support the rotating system.

In a conventional meter, the upper bearing consists of a guide pin (of hardened steel) moving inside a brass bush fitted to the shaft, and acts to provide lateral support at the top of the rotor assembly. The role of a lower bearing is far more important since, in addition to providing smooth rotation, it has to bear the weight or thrust of the rotating system. Even though the rotating system may not weigh much (typically not more than 20–30 g), the thrust borne by the bearing may be enormous owing to the very small area of contact. For this reason, the lower bearing comprises essentially a polished

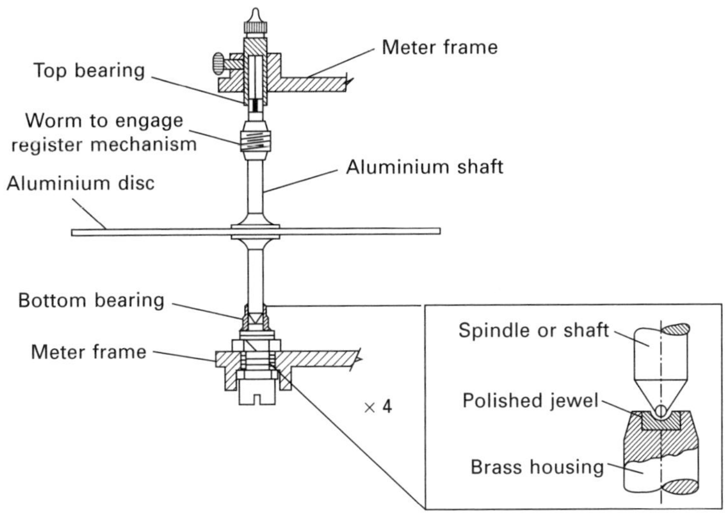

16.1 Rotating system of an energy meter with conventional or jewel bearing.

jewel cup and a hardened and polished pivot designed to give many years of trouble-free service. The schematic details of typical conventional bearings are shown in Fig. 16.1 in which the important design aspects of the lower bearing are brought out in the inset.

However, in spite of the best design and manufacturing efforts, the jewel bearings end to become 'sticky' due to continuous wear, and have an average life of about 10 years. 'Stickiness' may result in excessive friction, causing serious error in registering the energy consumption. The meter would then have to be disconnected, the bearing replaced and the meter recalibrated before it is put back into service.

A considerable increase in service life and reduction of maintenance cost by virtually eliminating wear can be achieved by using what can be termed as magnetic suspension bearings.

16.2.1 Magnetic suspension bearings

The essential feature of such bearings is the suspension or flotation of the meter rotor assembly by the magnetic force between two cylindrical permanent magnets, one of these being attached to the rotor shaft and free to rotate while the other is fixed to the stationary meter frame. The two magnets are never in contact and are separated by an air gap.

The suspension of a rotating system by means of magnetic force can be achieved in two possible ways: attraction or repulsion. In the 'attraction' type, a pair of cylindrical, axially magnetized permanent magnets is

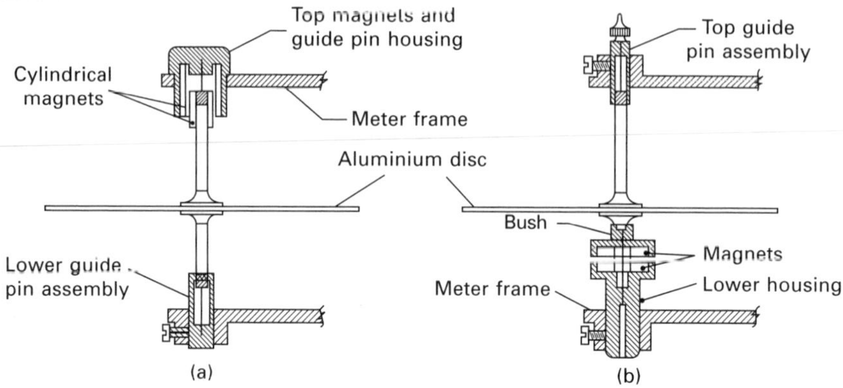

16.2 Magnetic bearings: (a) attraction type; (b) repulsion type.

concentrically mounted towards the topside of a rotor assembly, as shown in Fig. 16.2(a). The weight of the assembly is vertically supported by the force of attraction between the two magnets along their axial length. The lateral stability is provided by the guide pin arrangement at the top and bottom, in a somewhat similar way to that used at the top (as shown in Fig. 16.1).

In the 'repulsion' type, as illustrated in Fig. 16.2(b), the lower guide-pin assembly, shown in Fig. 16.2(a), assumes the top position while the weight of the rotating system is supported by the repulsive force between two 'washer-type' permanent magnets, magnetized on the inner and outer flat surfaces (as shown). The lateral stability is again provided by a suitable guide pin arrangement. The air gap between the magnets acts as a friction-free air cushion on which the rotor assembly floats. For many reasons, but mainly due to economy of manufacture, this type of suspension has been in common use.

This chapter describes the various design and constructional features of modern repulsion-type magnetic bearings. It includes experimental studies of frictional torque characteristics of a conventional (or jewel-type) bearing and a number of possible bearings using permanent magnets.

16.3 Design considerations

16.3.1 Theoretical

The force of repulsion between two similarly magnetized magnets is a function of flux density, B_g, in the air gap separating the magnets and their 'common' area of cross-section, a, normal to the direction of B_g. Thus, referring to Fig. 16.3, the repulsive force F(N) between magnets P and Q is given by:

$$F = \frac{B_g^2 a}{2\mu_0} \text{(N)} \qquad [16.1a]$$

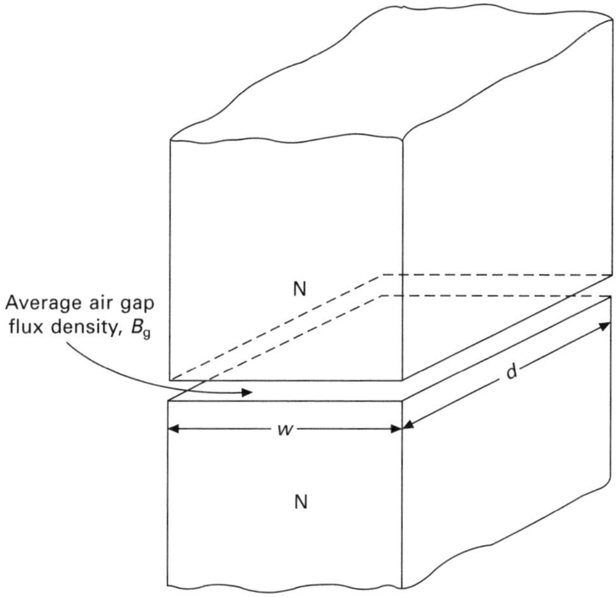

16.3 Similarly magnetized magnets held to produce a repulsive force.

where $a = w \times d$ (m^2), B_g is in tesla (T), and μ_0 is permeability of free space. Also,

$$F = \frac{B_g^2 a}{2\mu_0 g}(\text{kgf}) \qquad [16.1b]$$

Thus, two circular magnets of 10 mm diameter each, when placed opposite each other with only 0.1 T air gap flux density, would exert a force of about 32 g wt. – a force theoretically sufficient to support the weight of the disc assembly of an average single-phase energy meter, weighing about 20 g.

The axial length of the magnet can be worked out on the basis of energy contents in the air gap and the magnets. Assuming that the two have a common area (normal to B_g), it can be shown that the length of each of the magnets, l_m, is related to the desired air gap length, l_g, by:

$$l_m = 0.5\, l_g \times \lambda \qquad [16.2]$$

where λ is the leakage factor.

In practice, λ can vary from about 2.5 to 5 for permanent magnets usually employed for bearings and similar devices. Therefore, assuming a value of 4, the optimum length of each magnet would be twice the length of the air gap.

16.3.2 Practical aspects

It can be appreciated from Section 16.3.1 that the dimensions of magnets, for supporting a given rotor assembly of weight W_g are dependent on B_g, the air gap flux density, which, in turn, results from to the permanent magnet properties of the magnet material. The most important practical aspect of bearing design is, therefore, the selection of magnet material that would not only meet the technical requirements, but also be economical to keep the overall cost of the meter to a minimum. The careful selection of the material to be used for bearings must be followed by a close control on the manufacturing and magnetization requirements of the finished magnets to achieve the desired accuracy of operation.

16.3.3 Selection of magnet material

The main criteria for selecting material for the magnets are:

- The material should have an extremely high coercivity to counteract the risk of demagnetization by external fields.
- The material should possess high 'magnetic stability' for a period of 20–30 years. Normally, this quality is closely linked with the point above.
- The variation of air gap flux with ambient temperature should be negligible. This is a stringent requirement (since usually all the permanent magnet materials exhibit some dependence on temperature) and is very important to minimize vertical movement of the disc as a result of variation of air gap flux density. The positioning of the aluminium disc in the air gap of driving electromagnets is critical in an energy meter to maintain the accuracy, as any vertical displacement can give rise to serious errors.
- The material should possess as high 'stored magnetic energy' or $(BH)_{max}$, per unit volume as possible, to minimize the size.
- The material should also be of low density, to minimize the weight of the upper magnet attached to the spindle, in order to maintain low overall weight of the rotating system.

It is interesting to note that permanent magnet materials embodying almost all the above qualities are available – as a result of developmental efforts made since the 1960s towards improving their properties – but at a high cost. The high cost is generally attributable to the presence of rare and expensive alloying elements and the complicated processes necessary for the manufacture and heat treatment of these materials. For commercial and mass applications, therefore, a compromise is usually made between the desired properties and the cost of the material.

The commonly used magnetic materials and their important properties are compared in Table 16.1. Of these, the cheapest material, satisfying most

Table 16.1 Important properties of the common permanent magnet materials used for magnetic suspension bearings

Material	Typical values of				Remarks
	Remanence, B_r(T)	Coercivity H_c(kAm^{-1})	Stored magnetic energy, $(BH)_{max}$ (kJm^{-3})	Density, ρ (gcm^{-3})	
Barium ferrite	0.2	127.3	6.37	4.8	Cheapest; high thermal expansion coefficient
Anisotropic barium ferrite (Magnadur III) the Netherlands)	0.3	155	17.5	4.8	More expensive, preferable to barium ferrite
Alnico V (USA)	1.2	57.3	37.4	7.3	About 3 times as expensive as barium ferrite
Hycomax (UK)	0.9	66.1	25.45	7.25	Alnico V is more popular compared with
Alcomax III (UK)	1.25	53.3	40.6	7.35	Hycomax and Alcomax
Samarium cobalt, $SmCo_5$ (Recoma 20)	0.9	700	160	7.78	Latest; expensive; temperature thermal coefficient; –0.04%/°C, can operate; up to 250°C.

requirements, is barium ferrite or, better still, an isotropic barium ferrite, which is only slightly more expensive. This, therefore, finds application in most commercially manufactured energy meters provided with magnetic suspension bearings. A serious drawback of barium ferrite, however, is its high thermal expansion coefficient, which can result in up to 0.19% variation of air gap per degree centigrade change in ambient temperature. However, this deficiency can be overcome by fitting a suitable temperature compensator around the magnet, usually in the form of a thin collar of nickel.

The last material in Table 16.1, samarium cobalt, is the most modern, and one of the best, permanent magnet materials, possessing an extremely high coercivity and energy product, with an extensive range of applications. Its high cost is due to the presence of samarium and cobalt in the alloy; the high cost is well justified, however, in many applications on account of its very superior magnetic properties.

16.3.4 Manufacturing and magnetization requirements

These requirements can be broadly listed as follows:

- The magnets should be manufactured with close mechanical tolerances, both on the diameter and axial length giving absolutely flat and smooth surfaces. For this reason, the machinability of a given magnet material is a critical requirement. Usually, all permanent magnet materials are physically very hard and brittle, and their machining is always a problem. The magnets are, therefore, normally manufactured by casting, with the final finish imparted by grinding. This adds to the manufacturing cost.
- The finished magnets should be very carefully examined for their homogeneity and against possible internal deformities. This is important for achieving uniform magnetization throughout the magnet volume. Ultrasonic techniques are usually employed for such inspections.
- Finally, a careful magnetization of the magnet in appropriate directions is essential to impart near-perfect magnetic properties. Improper or non-uniform magnetization can not only result in undesirable, unbalanced forces during operation of the mechanism as a bearing (causing a 'drag' on the rotating system), but may also adversely affect the life of the magnets.

16.4 Mechanical requirements

The two most critical mechanical requirements, that must be satisfied in the bearings are as follows:

- The fixing of upper magnet to the disc concentric to the axis of rotation such that the surfaces of the magnets are perpendicular to the axis. The process should be simple and fast and yet very accurate. While many other techniques are possible, the one generally employed is die-casting the magnet at the bottom of aluminium disc, at the same time as the spindle or shaft is cast to the disc.
- The guide pin assembly used must provide the necessary lateral stability to the rotating system at its lower end by restricting horizontal movement of the shaft. The guide pin should be concentric with the axis of rotation and may introduce only negligible friction during rotation, particularly at slow speeds.

It is important to consider the following points:

- The material and finish of the pin and the bush. It is most common to use hardened and polished stainless-steel wire (of about 0.5 mm diameter) for the guide pin and a special variety of graphite for making the bush. The latter can also be manufactured from brass.

- The relative diameters of the guide pin and the bush.
- The depth of penetration of the guide pin inside the bush in the final assembly.

It is also desirable to adopt a simple design for quick and easy fitting and removal of the lower magnet assembly in the main frame of the meter, without sacrificing the concentricity of the rotating system.

16.5 Typical construction

A typical example of a repulsion-type bearing made from commercially available magnets is shown in Fig. 16.4. The material of the magnet was isotropic barium ferrite and the annuli were magnetized axially. A brass bush, details of which are shown in Fig. 16.4(b), was used to support the upper magnet and was fitted to the cut-away spindle of an existing disc assembly. For lateral stability, the guide pin that was normally used at the top was fitted in an inverted mode to the meter frame in the present case, through a brass sleeve supporting the lower magnet (as shown in Fig. 16.4).

In version II of the prototype, the various elements of the bearing and guide pin assembly were much the same as in Version I above, but the magnet pair M, M (Fig. 16.4(b)) was replaced by another pair. These magnets, although still made of barium ferrite, were characterized by better dimensional tolerance, improved finish and more uniform magnetization. The general quality of the assembly was also improved through close manufacturing tolerances and careful adjustment of internal bush diameter relative to guide pin diameter. As revealed by experimental results, these improvements yielded better performance of the bearing.

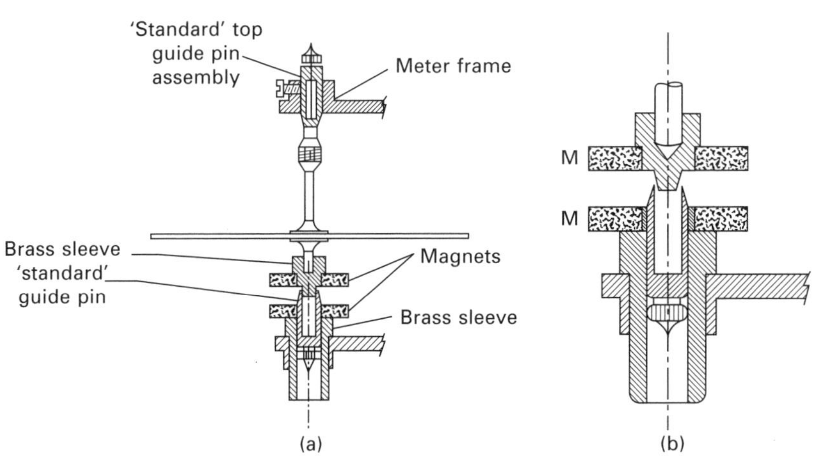

16.4 Indigenous magnetic bearing (version I): (a) the rotating system; (b) enlarged view of magnet assembly.

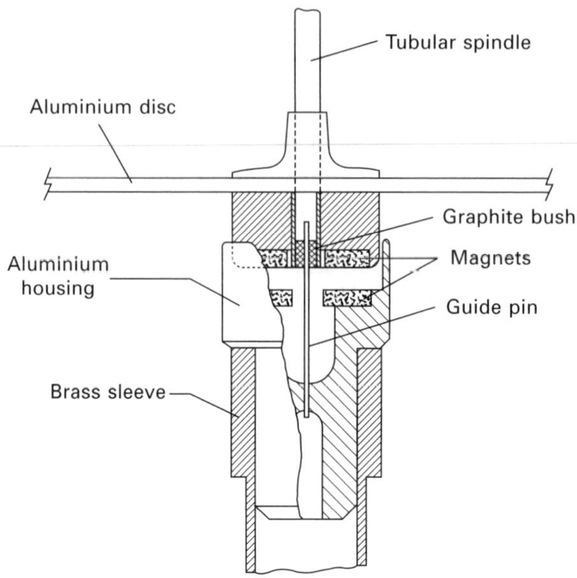

16.5 Enlarged view of the imported magnetic bearing assembly.

The details of a bearing assembly (from Japan) are illustrated in Fig. 16.5. The magnets are made from an isotropic barium ferrite. Since it has nearly three times the energy content/unit volume of ordinary barium ferrite (as shown in Table 16.1), the magnets were small in size and yet resulted in an air gap of about 1 mm while supporting a rotor of 90 mm diameter and of 22 g weight. It is to be noted that the upper magnet was die cast (using a special tin alloy) with an aluminium disc, and the enclosed shaft end was fitted with a tiny graphite bush. The guide pin was made of hardened stainless steel of 0.4 mm diameter and was highly polished. A useful feature of the assembly was the thin protective collar integral to the lower magnet housing, which completely surrounded the upper magnet, thus providing the necessary protection against shock and undue vibrations during transit.

16.6 Frictional torque studies

16.6.1 Testing procedures

The rotation of a disc in an energy meter at a given load is governed by the following three basic torques acting on the rotating system:

- the driving torque produced by the action of fluxes proportional to the supply voltage and the load current;
- the braking torque produced by the permanent magnet; and
- the frictional torque inherent to the driven assembly, the seat of which, as pointed out before, is mainly the lower bearing.

When an improvement is sought on the design of the lower assembly, it is therefore natural to study the frictional torque characteristics of the rotating system in isolation (that is, without being affected by the other two torques), to assess the performance.

16.6.2 'Speed-based' procedure

A simple procedure for evaluating relative frictional torque behaviour would be to impart a given transient impulse to the rotating system of the energy meter and count the number of revolutions made by the disc in a given time when the system is slowing down under the influence of its frictional torque.

Figure 16.6 shows the schematic of the procedure to be used. The aluminium disc D was supported at the top by the usual guide pin, and at the bottom by the given type of bearing. The disc was free to rotate in the air gap formed by the pair of electromagnets EMV and EMC, carrying currents proportional to the supply voltage and load current respectively. The supply was switched on by switch SW and, after the disc attained a steady speed, switched off (at time $t = 0$), causing the disc to slow. The number of revolutions made by the disc in a given time (e.g. 1 min) was recorded. The process was repeated with the various types of bearings fitted, in turn, to the same frame assembly. Under the prevailing atmospheric conditions, a comparison of number of revolutions made in the given time interval then provided a picture of relative performance of various bearings.

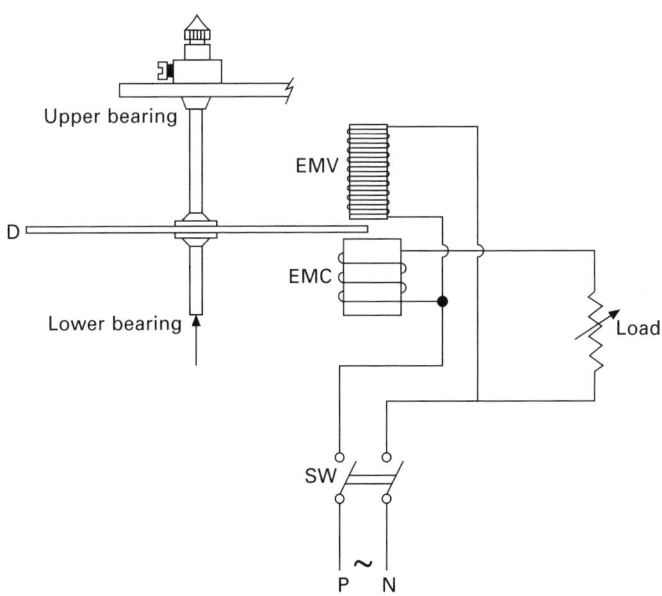

16.6 Schematic of 'speed-based' procedure.

16.6.3 'Time-and-revolutions-to-standstill' procedure

A method often used in practice is that of imparting a gentle push to the assembled disc in the meter without exciting any of the above electromagnets and recording the time and number of revolutions taken by the disc to come to rest. Then, if the disc completes n (fractions included) revolutions in time t before stopping, the frictional torque T_f, in Nm, of the bearing system is expressed by:

$$T_f = (Kn/t^2) \times 10^{-7} \qquad [16.3]$$

where

$$K = \text{a constant}$$

$$= 2\pi m r^2 \times 10^{-2} \qquad [16.4]$$

in which, m = mass of the rotor (g) and r = radius of the disc (mm). When interpreted in terms of identical initial angular speed ω of two systems under comparison, the above would mean that the system with smaller frictional torque would take much longer to come to standstill, as indicated in Fig. 16.7.

Since this method affords a quantitative assessment of the frictional torque for a rotating system, it was followed in all experimental studies.

16.6.4 Experimental results

The recorded values of n and t, and the calculated values of T_f, individually as well as the average for the type of assembly under test, are presented in Table 16.2. The reference case is taken to be that of a conventional jewel bearing which, as would be expected, gives the lowest frictional torque for a new, polished bearing assembly. The discussion of results follows in Section 16.7.

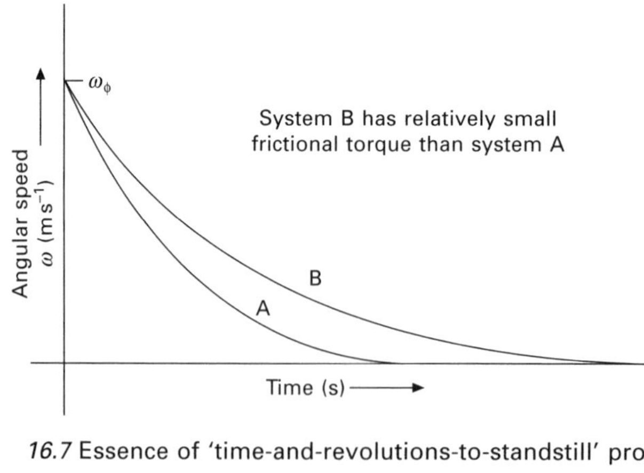

16.7 Essence of 'time-and-revolutions-to-standstill' procedure.

Table 16.2 Experimental results

A Conventional jewel bearings

Rotor diameter: 80 mm; rotor wt: 18 g; $K = 1809.55$

n (revolutions)	3.87	5.75	9	18.25	19.75
t (s)	36.5	46.5	52.5	80.5	82.0
$T_f \times 10^{-7}$ (N m)	5.3	4.8	5.9	5.1	5.3

T_f (av) $= 5.28 \times 10^{-7}$(N m)

B Magnetic bearing, using barium ferrite magnets of OD 20 mm, ID 6.5 mm, L 3 mm (Version I)

Rotor diameter: 80 mm; rotor wt: 23.16 g; $K = 2328.84$

n (revolutions)	2.0	3.0	8.75	17.0	29.0
t (s)	14.0	18.0	33.5	43.0	56.0
$T_f \times 10^{-7}$ (N m)	23.76	21.65	18.16	21.4	21.5

T_f (av) $= 21.3 \times 10^{-7}$ (N m)

C Magnetic bearing, using barium ferrite magnets of OD 15 mm, ID 8 mm, L 6 mm (Version II)

Rotor diameter: 80 mm; rotor wt: 23.66 g; $K = 2379.27$

n (revolutions)	4.5	6.25	10.38	17.25	21.5
t (s)	27.0	35.0	45.0	61.5	65.5
$T_f \times 10^{-7}$ (N m)	14.6	12.1	12.2	10.8	11.9

T_f (av) $= 12.3 \times 10^{-7}$ (N m)

D Japanese magnetic bearing, using anisotropic barium ferrite magnets of OD 10 mm, ID 2.5 mm, L 1.5 mm; 'standard' top guide pin, rotating inside graphite bush of diameter (enlarged to) 0.6 mm

Rotor diameter: 90 mm; rotor wt: 22.1 g; $K = 2811.88$

n (revolutions)	1.25	4.38	6.25	11.5	18.5
t (s)	23.5	42.5	53.0	63.0	80.0
$T_f \times 10^{-7}$ (N m)	6.36	6.8	6.26	8.14	8.13

T_f(av) $= 7.14 \times 10^{-7}$ (N m)

E Japanese magnetic bearing with reduced disc diameter, and spindle mechanism adapted to fit 'standard' top guide pin, rotating inside 'standard' brass bush

Rotor diameter: 80 mm; rotor wt: 19.4 g; $K = 1950.67$

n (revolutions)	2.0	4.63	6.5	12.88	19.75
t (s)	24.0	39.0	50.0	63.0	80.0
$T_f \times 10^{-7}$(N m)	6.77	5.93	5.07	6.33	6.02

T_f(av) $= 6.02 \times 10^{-7}$(N m)

Table 16.2 Continued

F Developed magnetic bearing rotor assembly with 'standard' top guide pin bottom brass housings fitted with Alnico V magnets, top guide pin used in inverted mode [as shown in Fig. 16.4 (b)]

Rotor diameter: 80 mm; rotor wt: 19.5 g; $K = 1963.83$

n (revolutions)	2.0	5.25	9.75	15.25	24.5
t (s)	15.0	22.5	33.0	39.0	52.0
$T_f \times 10^{-7}$(N m)	17.45	20.36	17.6	19.7	17.8

T_f (av) $= 18.6 \times 10^{-7}$ (N m)

G Japanese magnetic bearing rotor assembly with top guide pin rotating as in D and with original bottom guide pin mechanism, but both magnets replaced by barium ferrite magnets

Rotor diameter: 90 mm; rotor wt: 19.45 g; $K = 2474.71$

n (revolutions)	3.0	5.25	9.0	11.13	22.25
t (s)	27.5	31.0	44.0	50.0	69.0
$T_f \times 10^{-7}$ (N m)	9.8	13.5	11.5	11.01	11.56

T_f (av) $= 11.47 \times 10^{-7}$ (N m)

16.7 Conclusions and discussion

The average values of frictional torque in sets 'A' through 'C' (Table 16.2) are illustrative of varying degrees of frictional performance of energy-meter rotating systems. While it would seem that the conventional jewel bearing is perhaps the best, offering minimum frictional torque, it may be noted that the condition can deteriorate with time, first slowly and then rapidly, to result in very high frictional torque. In fact, this was observed in some sort of 'accelerated' friction test in which the rotor was rotated at a very fast speed continuously for more than an hour. The frictional torque, on re-evaluation using the above procedure, showed an increase 10%. The frictional torque in a magnet bearing, however, would remain uniform over a much longer period.

A marked improvement in the value of frictional torque in version II emphasizes the significance of the quality of magnet used, design optimization of bottom guide pin mechanism, and concentricity and alignment of the rotating system. In the overall assembly, the depth of the guide pin inside is also important.

It must be remembered that the size of the magnets used in such prototypes was very much restricted by the very limited availability of such magnets in the market. Thus, the magnets were of relatively large size, resulting in rather heavy rotating systems (affecting the inertia and hence the frictional torque) and unnecessary long air gaps (of the order of 5 mm). Given suitable magnets, the performance of the developed bearings might have been appreciably better.

A comparison of results from sets 'D' and 'E' shows an improvement in frictional torque by the reduction of disc diameter and weight. It is also clear that the selection of materials and design of the top guide pin assembly have a negligible effect on the bearing performance although, at least qualitatively, it would be desirable to use a well-polished guide pin rotating inside a non-lubricated graphite bush of appropriate inside diameter.

The effects of modifications in the bearing (from Japan), by way of interchanging magnets or guide pin assemblies, are reflected in the results of sets 'F' and 'G'. These demonstrate that it may be relatively easy to replace magnets from Japanese bearings by locally available ones; in the beginning, only the requirements of magnetic force of repulsion need be considered, although in the long run other factors such as magnetic and temperature stability would also have to be considered. By comparison, the design and application of a particular guide pin mechanism for lateral support at the bottom have more bearing on the frictional torque. While much would depend on the materials used for making the guide pin and the bush, the assembly and alignment of these parts is also significant. It is to be noted that ordinary barium ferrite magnets were used in case 'G'. With a proper selection and use of good quality magnets, the results could have been still better.

The study given in this chapter, while giving a good picture of the types and application of magnetic bearings in energy meters, shows what needs to be done to develop and manufacture such bearings of good quality. Notwithstanding the fact that the frictional torque of the developed magnetic bearing is more than twice that of a conventional jewel bearing, it may be pointed out that the value is still only about one-third of the minimum starting torque for which the meters are designed. With judicial selection of key materials and components and extensive developmental efforts, it should be possible to bring about a considerable improvement in the performance of magnetic bearings.

16.8 Bibliography

1 Shotter, G.F. and Tagg, G.F. *Induction Type Integrating Meters*, Pitman & Sons, London.

2 Trekeli, H.F., Mendelssohn, L.I. and Wright, J.H., 'Magnetic Suspension of the Watt Hour Meter Rotor', *Trans. AIEE*, **57**, 1180, 1948.

3 *Magnetic Suspension Bearings*, GEC Publication No. M5-002E, UK.

4 Glyde, T., 'Meters with Magnetic Suspension', *International Conference on Metering and Apparatus for Modern Electricity Supply Tariffs*, London, 1967.

5 Lovegrove, H.J., 'Magnetic Bearings for Watthour Meters' *International Conference on Metering and Apparatus for Modern Electricity Supply Tariffs*, London, 1967.

6 Bhargava, S.C. and Prashad, H., 'Magnetic Suspension Bearings for A.C. Energy Meters', *BHEL J.*, **6**(2), 12–21, 1985.

17
A new generation of rolling-element bearing with an outline of its performance advantages

17.1 A general review

This chapter deals with the concept, development and performance evaluation/investigation with an outline of advantages of a new-generation double decker high-precision rolling-element bearing (DDHPB). The chapter brings out the basic design concept of DDHPB and establishes that the use of two pairs of rolling-element series – one riding on the other and separated by the intermediate rotating race between them – facilitates using the bearing for moderate speed and may provide higher-speed applications with better performance and lower losses. Theoretical analysis of DDHPB has shown that the reduction in speed of the intermediate rotating race relative to that of the shaft speed reduces the operating temperature and frictional torque of the DDHPB bearing compared with the equivalent conventional bearing of the same shaft diameter. This has also been established by experimental investigations.

17.2 Introduction

Fatigue and surface distress usually describe the limits for reliable operation of a rolling-element bearing. Under various operating conditions, the bearings have different life spans and, in general, bearing failure is caused by surface distress and is indicated by temperature rise. This leads to bearing damage in due course, resulting in micropitting, smearing, and indentation and plastic deformation, besides surface corrosion. Depending on the application demands, bearings work under high speed, high temperature and heavily stressed conditions. This calls for technical innovations covering materials, change in bearing design configuration, use of modified surface coatings and lubricants.

Keeping the above requirement in view, Anderson proposed a unique bearing concept with a basic design modification for high speeds of the order dn = 3 000 000 mm rpm.[1] Anderson's bearing concept, however, combined one rolling and one hydrodynamic sliding bearing. To meet the same needs, the double-decker rolling-element bearing concept has been developed. This study reports the principle of operation, theoretical background and experimental investigation to highlight the improved performance and advantages of the new-generation DDHPB for moderate speed applications, and compares its performance with that of the equivalent conventional bearing. The bearing may also demonstrate better performance for high-speed applications.

17.3 Basic concept and principle of operation of DDHPB

The basic concept of the DDHPB is to use two pairs of rolling-element series, one riding on the other and separated by the intermediate rotating race in between them. The intermediate race acts as an inner race of a second series of rolling elements (secondary) and as an outer race for the first series of rolling elements (parent); however, the outer race of the second series of rolling elements is mounted in the housing, whereas the inner race of the first series of rolling elements is fitted on the rotating shaft like a conventional bearing. Figure 17.1 shows the basic configuration of DDHPB design.

The inner race of the first series of rolling elements of DDHPB rotates at the shaft speed; however, the intermediate race rotates at lower speed than the inner race because of frictional forces, kinematics and configuration of the rolling elements, and slip phenomenon occurring between the races and the rolling elements. This has been established both by theoretical and experimental investigations.

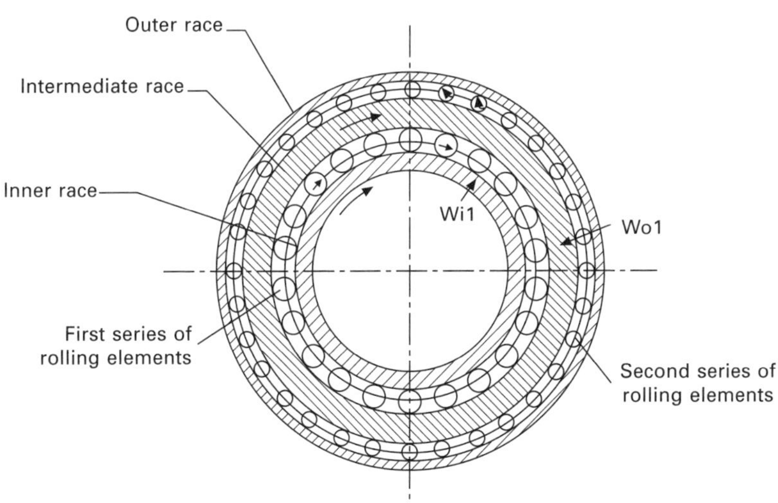

17.1 Schematic of double-decker high-precision ball bearing (DDHPB).

The rotational speed of the intermediate race is determined assuming that the driving torque in both rolling-element rows is equal. Equality can be achieved with or without slip in either row of rolling elements. Taking the case of no slip in either row, it is reasonable to hypothesize that the driving torque for either of the bearing rows will increase with increasing relative velocity between its rings, once any starting torque peak has been overcome. As the absolute speed of the intermediate ring is increased, its relative velocity with respect to the stationary outer ring will increase, but its relative velocity with respect to the rotating inner ring will decrease. Therefore, a stable equilibrium intermediate ring speed is generally achievable. Its magnitude depends on specific torque/speed function prevailing in each of the two rolling-element rows, and these functions are not identical in general.

If the existence of slip is admitted, then the torque equilibrium problem becomes more complex, but the same principle applies. The mathematical treatment needs to proceed on the basis of these facts, and will then lead to the conclusion that there is no geometrically determined unique non-slip centre.

17.4 Theoretical analysis

17.4.1 Effect of cage and roller slip on the rotation of intermediate race of the DDHPB

In most rolling bearing applications, operating conditions are such that the cage and rolling-elements motions are essentially epicyclical. However, in some situations of high speeds and light loads, several investigators have reported cage slip, in which cage and rolling elements assembly travel at speeds lower than that predicted from epicyclical considerations.[2,3] Smith reported considerable cage slip in the main shaft bearings operating at high speeds under light loads.[4] Dowson and Higginson[5] suggested that considerable cage and roller slip could be expected in lightly loaded bearings operating at moderate speeds; whereas under heavier loads when the conditions would be elastrohydrodynamic, the rolling element motion would be essentially epicyclical. However, Harries,[6] by applying analytical methods based on elastohydrodynamic considerations, has shown that significant departure from epicyclical motion can exist in heavily loaded bearings operating even under elastohydrodynamic condition.

The effect of cage and roller slip on the measured defect frequency response of rolling-element bearing, has been reported,[7] which indicates that the negative cage and roller slip is predominant at moderate speed under 'no-load' and 'load' conditions. Positive slip is predominant at high speed under 'no-load' operation.

In general, the percentage of negative cage slip is found to be more than the corresponding rolling-element slip and the percentage of positive cage slip is less than corresponding rolling-element slip.[7] Negative cage and roller slip indicates faster motion of cage and rolling element than the epicyclical value, whereas positive cage and rolling slip shows the slower motion of cage and rolling element than its epicyclical values as determined by the kinematics of the bearings.

Increase in speed under 'no-load' and moderate 'load' generally increases the oil film thickness between the inner ring and rolling elements, causing a tangential driving force, which is inversely proportional to the film thickness and tends to decrease it. Thus, the rolling elements rotate more slowly than their epicyclical value. However, the radial outward movements of the rolling elements due to an increase in film thickness results in a larger frictional drag force between the first series of rolling elements and intermediate race (Fig. 17.1). Since the intermediate race is a floating race, it transmits modulated frictional drag to the outer race of the secondary series through rolling elements. Thus, the intermediate race rotates more slowly, owing to the decrease in tangential drive force and modulated friction drag forces, than its epicyclical speed derived from the kinematics considerations.

Under moderate and high speed under high 'load' conditions, film thickness between the inner race and rolling elements does not increase considerably; so the decrease in tangential force is not predominant. Thus, the rolling elements tend to rotate faster than the epicyclical value. Also, there occur fewer friction drag forces between the first series of rolling elements and the intermediate race due to comparatively less radial outward movement because of reduced film thickness. This tends to increase the speed of the intermediate race, but, owing to close interaction with the second series of rolling elements, where modulation of the film thickness due to distribution of load takes place, altering the film thickness, driving force and frictional drag cause the intermediate race to rotate at much lower speed than the epicyclical value. Thus, the speed of the intermediate race is modified because of the positive and negative slip phenomenon and it rotates at a speed less than the epicyclical value determined by kinematics consideration.

At the moderate speed and high load application, there will be further reduction of intermediate race speed in both the first and the second series of rolling elements; so, considering the slip factor under these conditions, angular speed of the intermediate race can also be determined analytically.

17.4.2 Determination of the rotational speed of intermediate rotating race of the DDHPB (by driving torque equilibrium)

A reasonable estimate of the frictional torque of a given rolling bearing under moderate load and speed conditions is the sum of the load torque, viscous frictional torque and rolling end-flange frictional torque. For determination of speed of the intermediate rotating race by driving torque equilibrium, the viscous frictional torque is the most significant since both the series of rolling-elements are under the same load.

The viscous frictional torque of the first and the second series of rolling elements (M_p and M_s in kgmm) can be determined by the empirical formulae given by Palmgren[8] for $\nu n > 2000$, considering f_o and ν identical for both the series of rolling elements. These are determined as:

$$M_p = f_0 \times 10^{-8} (\nu n)^{2/3} d_{e_1}^3 \qquad [17.1]$$

and

$$M_s = f_0 \times 10^{-8} (\nu n_i)^{2/3} d_{e_2}^3 \qquad [17.2]$$

The rotational speed of the intermediate race (n_i) is determined on equating M_p and M_s, assuming that the driving torque under stable conditions in both the rolling-element rows is equal under no slip conditions, and is given as:

$$n_i = n(d_{e_1} / d_{e_2})^{4.5} \qquad [17.3]$$

17.5 Theory behind performance evaluation of bearings

Knowledge of bearing temperature is essential to determine the elastohydrodynamic thickness of the film, which is a significant factor for determining the lubricant specifications and in the predetermination of the frictional characteristics and thus generation of heat in a bearing. Heat generated in a bearing depends on the operating parameters, viscosity of lubricant and bearing design.[8,9] Of these variables, normally viscosity of lubricant is most affected by temperature. The parameters of bearing design/geometry, i.e. average diameter, dimensions of rolling elements, number of rolling elements (Z) and angle of outer circular paths, are summarized as a geometry factor of the bearing, which in turn is considered as a measure of the friction and heat generated in the bearing and hence the rise of its temperature. This indicates that bearing temperature, more precisely the outer race temperature, is the best indicator of the bearing performance.

In short, energy losses in the bearing are proportional to the frictional torque and angular velocity. This can be expressed mathematically as follows:[10,11]

$$P = MW \qquad [17.4]$$

The energy loss in the bearing gives rise to a temperature difference ΔT between the bearing and the ambient temperature as follows:

$$\Delta T = KP \qquad [17.5]$$

By combining equations (17.4) and (17.5), it follows that:

$$\Delta T = KMW \qquad [17.6]$$

From this, it is obvious that the frictional torque M is the main variable that can be influenced with regard to the generation of heat and which, in turn, influences the performance of bearings at moderate to high speeds. The total friction of a bearing is the sum of rolling, sliding and lubricant friction, which is the measure of resistance thc bearing exerts against its movement. This resistance represents a torque and is termed frictional torque. It is evident that the temperature rise of a rolling-element bearing depends on the frictional torque, angular velocity and bearing design factors. The temperature rise of a bearing gives performance characteristics of different bearings under identical conditions of operation (Equation 17.6).

17.5.1 Coefficient of friction

The difference in frictional behaviour between the various rolling-element bearings is recognized on considering the coefficient of friction (μ) instead of frictional torque, which is given as:

$$\mu = \frac{M}{W_w d/2} \qquad [17.7]$$

The ratio of coefficient of friction of DDHPB and equivalent conventional rolling-element bearing under identical operating condition is the same as the ratio of the frictional torque (Equations 17.6 and 17.7):

$$\frac{\mu_1}{\mu_2} = \frac{M_1}{M_2} = \frac{\Delta T_1}{\Delta T_2} \qquad [17.8]$$

17.6 Design and test conditions of DDHPB

A DDHPB ball bearing of inner diameter 50 mm and outside diameter of 100 mm with an intermediate rotating race of 80 mm diameter has been developed by BHEL. Two series of balls, one inside and the other outside the intermediate rotating race, have been provided to fit within the outside diameter of 100 mm. Details of the DDHPB developed are as follows:

- $d = 50$ mm
- $d_1 = 5.48$ mm
- $d_{e_1} = 65$ mm
- $d_2 = 3.54$ mm
- $d_{e_2} = 90$ mm
- B = 16 mm
- Number of rolling elements in the first series = 14
- Number of rolling elements in the second series = 27

The DDHPB developed is shown in Fig. 17.2. It has been tested at the speeds of 500, 1000, 1500, 2000 and 2500 rpm on the bearing test rig (Fig. 17.3) at 500, 1000, 1500, 2000 and 2500 N of radial loads. The equivalent conventional deep-groove ball bearing of 50 mm diameter type 6010 has also been tested under similar load and speed conditions, and its performance was compared

17.2 Double-decker high-precision ball bearing (50 mm bore).

17.3 Bearing test rig.

with that of the developed DDHPB bearing. Similarly, other sets of DDHPB rolling-element bearings have been developed, and their performance characteristics compared with the conventional bearings.

A small window on the housing of the DDHPB, $25 \times 40\,mm^2$ size, was made in order to look through the intermediate floating race by use of a fibre optic probe, and the speed was precisely monitored using DVF-3 and a non-contact optical probe under different speeds of operation at various load patterns on the DDHPB.

17.7 Bearing test set-up and experimental details for testing DDHPB vis-à-vis conventional bearings

The conventional bearings and DDHPBs were tested separately on the bearing test rig shown in Fig. 17.3. The test rig is driven by a 3 horse power, variable-speed thyristor-controlled DC motor. The test rig shaft is mounted on two support bearings: a self-aligning ball bearing and the other end support the motor end of the test shaft by a pressure-fed journal bearing. The support bearings pedestals are fitted to a T-slot test bed. The test bearings (DDHPB and conventional) are mounted freely on the shaft end independently in the different plumber blocks for the performance evaluation. The DDHPB and conventional test bearings are loaded through the plumber block by means of a screw arrangement through a threaded rod attached to the block during test. Loading is provided by means of a lock-tightening arrangement, and monitored by a load cell attached to a plate, which is connected to a digital indicator in the instrument panel.

Thermocouple holes are made on each plumber block to monitor the temperature of the test bearing outer race. Also, accelerometers of 4371 B & K type along with charge amplifiers are used to monitor the vibration levels and for spectrum analysis from the bearing housing. A shock pulse meter is also used periodically to monitor the test bearing condition.

17.8 Data deduction

A rise in temperature above ambient has been monitored for the DDHPB and conventional bearings under various parameters of operation. Table 17.1 presents the data. Based on temperature rise and operating speed, the frictional torque characteristics of DDHPB and convention bearings are compared using equation (17.8) as shown in Table 17.2. The speeds of the intermediate race under various shaft speeds varying from 500 to 2500 rpm with the increment of 250 rpm at different radial loads of 500, 1000, 1500, 2000 and 2500 N were monitored precisely, and are presented in Table 17.3. These results were compared with the theoretically derived speed of the intermediate

Table 17.1 Experimental determination of rise in temperature of outer race of the DDHPB and equivalent conventional bearings under various conditions of operation

Speed (rpm)	Rise in temperature (°C)						Temperature stabilized above ambient (°C)						Difference in temperature rise (°C)					
	DDHPB						Conventional bearing						DDHPB-Conventional bearing					
	Load (N)																	
	No. load	500	1000	1500	2000	2500	No load	500	1000	1500	2000	2500	No load	500	1000	1500	2000	2500
500	3.5	6.3	–	–	–	–	13.3	15.1	8.6	9.9	10.5	12.3	–9.8	–8.8		–	–	–
1000	4.5	8.4	8.2	10.1	12.8	11.9	–	21.1	12.4	16.1	16.4	20.0	–	–12.7	–4.2	–6	–3.6	–8.1
1500	–	15.8	12.5	19.3	17.3	13.7	–	29.6	14.2	21.3	22.7	27.6	–	–13.8	–1.7	–2	–5.4	–13.9
2000	–	22.9	23.5	25.6	27.3	24.0	–	31.8	25.2	27.0	29.3	35.4	–	–8.9	–1.7	–1.4	–2.0	–11.4
2500	–	26.5	26.7	30.4	30.0	24.6	–	33.0	32.0	34.0	35.5	40.8	–	–6.5	–5.3	–3.6	–5.5	–16.2

Table 17.2 Ratio of frictional torque of the DDHPB and equivalent conventional bearing under various conditions of operation

Speed (rpm)	Load (N)					
	No load	500	1000	1500	2000	2500
500	0.26	0.42*	–	–	–	–
1000	–	0.39*	0.66	0.63	0.78	0.60
1500	–	0.53	0.88	0.91	0.76	0.50
2000	–	0.72	0.93	0.95	0.93	0.68
2500	–	0.80	0.83	0.89	0.85	0.60

*Light load and low-speed values ignored for analysis.

Table 17.3 Experimental evaluation of the speed of intermediate race of the DDHPB under various shaft speeds and load conditions

Shaft speed	500 N		1000 N		1500 N		2000 N		2500 N		No load	
	Speed of intermediate race and its percentage with respect to shaft speed											
(rpm)	rpm	%	rpm	%	rpm	%	rpm	%	rpm	%	rpm	%
500	128	26	114	23	123	25	116	23	78	16	126	25
750	210	28	180	24	219	29	180	24	122	16	210	28
1000	258	26	222	23	247	25	256	26	156	16	243	24
1250	332	27	277	24	305	24	324	26	275	22	292	24
1500	345	23	312	23	343	23	370	25	344	23	–	–
1750	385	22	360	21	380	22	409	23	363	21	–	–
2000	397	20	421	21	432	22	439	22	460	23	–	–
2250	409	18	431	19	431	19	443	20	443	20	–	–
2500	451	18	464	19	479	19	477	19	442	18	–	–

1. Theoretical speed of the intermediate race without slip is 23% of the shaft speed as calculated by driving torque equilibrium (Equation 17.3).
2. Considering the negative and positive slip phenomena, the speed of the intermediate race with respect to shaft speed varies between 16 and 29%, as determined experimentally.

race obtained by the approach of driving torque equilibrium (Equation 17.3) and values are shown in Table 17.3.

17.9 Results and discussion

17.9.1 Temperature rise and frictional torque characteristics of DDHPB and conventional bearings

The temperature rise at which a rolling-element bearing operates is a function of many variables, i.e. load, speed, frictional torque, lubricant characteristics, bearing mounting and environmental conditions. Under steady-state conditions, the temperature rise of the bearings indicates the relative ability of one bearing system to another. Under identical conditions of operation, the outer race temperature rise of DDHPB is lower than the equivalent conventional bearing. The difference in temperature rise varies from 1.7 to 16.2 °C (Table 17.1), which is 7.0–40% less than the rise in temperature of the equivalent conventional bearing. The difference in rise in temperature of the DDHPB and equivalent conventional bearing (8.1–16.2 °C) at various operating speeds at 2500 N load is much higher than at a load of 1000 N (Table 17.1). This may be attributed to a positive slip phenomenon, which reduces the speed of the intermediate race to approximately 16–29% (Table 17.3). The rise in temperature and frictional torque of DDHPB bearing is lower than that of the equivalent conventional bearing in all the parameters of operation, but the variation is non-linear (Tables 17.1 and 17.2). This is attributed to the inconsistent slip phenomenon occurring during interaction of the intermediate race with the first and second series of rolling-elements. Also, the uniform redistribution of load and reduction in load at apex position (p) may contribute to the lower temperature rise of a DDHPB.

The ratio of frictional torque and coefficient of friction of DDHPB versus the equivalent conventional bearing at 2500 N varies from 0.50 to 0.68 as against 0.66 to 0.93 at 1000 N at different operating speeds (Table 17.2). This makes it evident that there exists a considerable difference between the performance of DDHPB and the equivalent conventional bearing although all factors including separator material, conformity, surface roughness, lubrication, etc. were identical in both the bearings.

17.9.2 Theoretical and experimental evaluation of speed of the intermediate rotating race of the DDHPB

The experimental results indicate that the speed of the intermediate race varies from approximately 16% to 29% under different operating parameters with respect to that of rotating shaft speed (Table 17.3). The experimental measured speed of the intermediate race at various operating conditions (Table 17.3) may match the theoretical speed if obtained under the effect of slip phenomenon. The speed of the intermediate race determined by the driving torque equilibrium using equation (17.3) is 23% of that of the shaft speed and matches more closely that of the experimental values (16–29%) under various operating conditions, as shown in Table 17.3.

The reduction in the speed of the intermediate race speed improves the performance characteristics of the DDHPB over those of the equivalent conventional bearing under the identical shaft speed. This is indicated by a lower temperature rise and frictional torque/coefficient of friction of the

DDHPB as compared to the equivalent conventional bearing (Tables 17.1 and 17.2).

17.10 Brief summary of the published research on DDHPB

Experimentally stiffness and damping properties of DDHPBs were compared with those of conventional rolling-element bearings. It has been reported that the stiffness and damping characteristics of DDHPB are better than those of the conventional bearings.[14]

Experimental investigations have been carried out to study the stress distribution on the surface of the outer race of the bearing by strain measurement under static conditions to evaluate the performance and relative comparison of the life of the DDHPB and the conventional bearings.[15] Furthermore, it has been established from the pattern of cyclic stresses that the lower frequency of the stress cycles under load and narrower range of the variation of cyclic stress in each rotation result in lower indentation of the rotating elements. This gives a higher fatigue life of DDHPB compared with the conventional bearings of the same bore and lower/same outer diameter as that of the DDHPB under identical moderate operating conditions.[16] Also, the power loss characteristic of DDHPB and conventional bearings under various operating parameters has been established experimentally to bring out the energy saving by DDHPB vis-à-vis conventional bearings.[17] Besides this, a theoretical model to determine radial deflection, indentation on the track surfaces and stiffness of DDHPB, and comparison with that of different equivalent deep groove conventional bearings of the same bore has been developed.[18] Furthermore, theoretical analysis of axial deflection of DDHPB under the influence of axial loads and its comparison with that of the conventional deep groove and angular contact ball bearings of the same bore and outer diameter has been done.[19]

Analysis has been carried out[20] of centrifugal forces acting on the elements of DDHPB and conventional bearings under identical operating conditions so as to ascertain analytically the results of experimental investigations, which were used to establish the performance characteristics of these bearings as reported in references 12 to 19.

17.11 Conclusions

From the above investigations the following conclusions are drawn:[12,13]

- A decrease in frictional torque/coefficient of friction and lower temperature rise in a rolling-element bearing is achieved by using equivalent double-decker high-precision bearing (DDHPB).

- The maximum reduction in temperature rise and frictional torque/coefficient of friction, to the extent of 50%, as compared with that of the equivalent conventional bearing, can be achieved by DDHPB under certain operating conditions.
- Reduction in speed of the intermediate race up to 16–29% of the shaft speed, as compared with that of the conventional bearing, is achieved experimentally by the DDHPB.
- Theoretical analysis of the speed of the intermediate race of the DDHPB is 23%, obtained by the driving torque equilibrium approach; this matches the experimental values closely.

In all those applications where space restriction in the axial direction plays a significant role, as in motors/machine tool bearings at moderate or perhaps high speed, the application of DDHPB bearings will be an asset, in addition to the better characteristics of these bearings.[14–20]

17.12 References

1. Anderson, W.J., *Tribology for Aerospace Application*, SKF, pp. 526–528, 1973.
2. Hamrock, B.J. and Dowson, D., *Ball Bearing Lubrication*, John Wiley and Sons, 1981.
3. Boness, R.J., 'Cage and Roller Slip in High Speed Roller Bearings', *J. Mech. Sci.*, **II**, 2, 1969.
4. Smith, C.F., 'Some Aspects of the Performance of High Speed Lightly Loaded Cylindrical Roller Bearing', *Proc. Inst. Mech. Eng.*, **176**(227), 566, 1962.
5. Dowson, D. and Higginson, G.R., 'Theory of Roller Bearing Lubrication and Deformation', *Lubrication and Wear Convention* (Institution of Mechanical Engineers), London, p. 216, 1963.
6. Harris, T.A., 'An Analytical Method to Predict Skidding in High Speed Roller Bearings', *ASLE Trans.*, **9**(3), 229–241, 1966.
7. Prashad, H., 'The Effect of Cage and Roller Slip on the Measured Defect Frequency Response of Rolling Element Bearing', *ASLE Trans*, **30**(3), 360–367, 1986.
8. Tedric, H.A., *Rolling Bearing Analysis*, John Wiley and Sons, Inc., New York, 1966.
9. Houghton, P.S., *Ball and Roller Bearings*, Applied Science Publishers Ltd, London, 1976.
10. Burnaby Lautier, E. and Jutpass, E.R.C., 'Test Rig for Axle Box Bearings for High Speed Trains', *Ball Bearing J.*, **141**, 1978.
11. Beagley, T.M., 'SKF "THISBE" Test Rig', SKF Report No. NL 777515, 1978.
12. Prashad, H., 'A New Generation Double Decker High Precision Rolling-element Bearings – Concept, Development and Investigations', *Tribol. Trans.*, **44**(2), 203–208, April 2001.
13. Prashad, H., 'New Generation Rolling-element Bearings – An Investigation', *BHEL J.*, **23**(2), 43–56, October 2002.
14. Prashad, H., 'Relative Comparison of Stiffness and Damping Properties of Double Decker High Precision and Conventional Rolling-element Bearing', *Tribol. Int.*, **35**, 265–269, 2002.

15 Prashad, H., 'Experimental Evaluation of the Stress Distribution on the Outer Surface of the Outer Race of Conventional and Double Decker High Precision Bearing', *Tribo. Test, J.*, **9**(3), 249–260, March 2003.
16 Prashad, H., 'Pattern of Cyclic Stresses on Outer Race of Conventional Bearings vis-à-vis Double Decker High Precision Bearing – An Investigation', *BHEL J.*, **24**(2) 38–47, June 2003.
17 Prashad, H., 'Energy Efficient Bearings – An Investigation', *J. Lub. Eng.*, **59**(6) 17–22, June 2003.
18 Prashad, H., 'A Theoretical Approach to Evaluating the Performance Characteristics of Double Decker High-Precision Bearings', *Tribotest, J.*, **10**(3), 251–263, March 2004.
19 Prashad, H., 'An Analysis of Axial Deflection of Double Decker High Precision Bearings vis-à-vis Conventional Ball Bearings', *J. Lub. Eng.*, 2006.
20 Prashad, H., 'Centrifugal Forces on Double Decker High Precision and Conventional Ball Bearings', *J. Inst. Eng.* (I), **86**, 109–114, July 2005.

17.13 Nomenclature

B width of bearing
d bore diameter
d_1 diameter of first series of rolling element
d_2 diameter of second series of rolling element
d_{e_1} pitch diameter of parent bearing formed by first series of rolling elements
d_{e_2} pitch diameter of secondary bearing formed by second series of rolling elements
f_o factor depending on bearing design and lubrication method
K constant depending on bearing speed and design factors
M friction torque of bearing
M_1 frictional torque of DDHPB
M_2 frictional torque of equivalent rolling-element bearing
M_p frictional torque of parent bearing formed by first series of rolling-elements
M_s frictional torque of secondary bearing formed by second series of rolling-elements
n rpm of shaft
n_i rpm of intermediate race
p apex loading
P energy loss in bearing
W angular speed of bearing
W_i angular speed of inner race of parent bearing
W_o angular speed of outer race of parent bearing
W_w bearing load
Z number of rolling elements
ΔT rise in bearing temperature

ΔT_1	rise in temperature of DDHPB
ΔT_2	rise in temperature of conventional bearing
μ_1	coefficient of friction in DDHPB
μ_2	coefficient of friction in equivalent rolling-element bearing
μ	coefficient of friction
ν	lubricant viscosity in centistoke ($mm^2 s^{-1}$)

Index